Trout and Salmon: Ecol
and Rehabili

C000151204

Trout and Salmon: Ecology, Conservation and Rehabilitation

D.T. Crisp

BSc, PhD, DSc, Chartered Biologist
Independent Fishing Consultant
FI Biol.

Fishing News Books
An imprint of Blackwell Science

b

Blackwell
Science

© 2000 Fishing News Books
a division of Blackwell Science Ltd
Editorial Offices:
Osney Mead, Oxford OX2 0EL
25 John Street, London WC1N 2BL
23 Ainslie Place, Edinburgh EH3 6AJ
350 Main Street, Malden
  MA 02148 5018, USA
54 University Street, Carlton
  Victoria 3053, Australia
10, rue Casimir Delavigne
  75006 Paris, France

Other Editorial Offices:

Blackwell Wissenschafts-Verlag GmbH
Kurfürstendamm 57
10707 Berlin, Germany

Blackwell Science KK
MG Kodenmacho Building
7–10 Kodenmacho Nihombashi
Chuo-ku, Tokyo 104, Japan

First published 2000

Set in 10/13pt Times
by DP Photosetting, Aylesbury, Bucks
Printed and bound in Great Britain at
the University Press, Cambridge

The Blackwell Science logo is a trade mark of
Blackwell Science Ltd, registered at the United
Kingdom Trade Marks Registry

DISTRIBUTORS

Marston Book Services Ltd
PO Box 269
Abingdon
Oxon OX14 4YN
(*Orders:* Tel: 01865 206206
       Fax: 01865 721205
       Telex: 83355 MEDBOK G)

USA
Blackwell Science, Inc.
Commerce Place
350 Main Street
Malden, MA 02148 5018
(*Orders:* Tel: 800 759 6102
       781 388 8250
       Fax: 781 388 8255)

Canada
Login Brothers Book Company
324 Saulteaux Crescent
Winnipeg, Manitoba R3J 3T2
(*Orders:* Tel: 204 837 2987
       Fax: 204 837 3116)

Australia
Blackwell Science Pty Ltd
54 University Street
Carlton, Victoria 3053
(*Orders:* Tel: 03 9347 0300
       Fax: 03 9347 5001)

A catalogue record for this title is
available from the British Library

ISBN 0-85238-256-1

Library of Congress
Cataloging-in-Publication Data
Crisp, D.T. (D. Trevor)
  Trout and salmon/D.T. Crisp.
    p. cm.
    Includes bibliographical reference (p.   ).
    ISBN 0-85238-256-1
    1. Brown trout.   2. Atlantic salmon.
  I. Title.
  QL638.S2C75   1999
  597.5'6 – dc21   99-32068
                                        CIP

For further information on
Fishing News Books, visit our website:
http://www.blacksci.co.uk/fnb/

For Diane, Fiona and David

Contents

Introduction

The main subjects of this book are the two species of the genus *Salmo*, namely the 'European' trout (*Salmo trutta* L.) and the Atlantic salmon (*S. salar* L.) particularly during the parts of their lives that are spent in flowing fresh water. In the Pacific region there is a larger and economically more important genus of trout and salmon known as *Oncorhynchus*. As the genera *Salmo* and *Oncorhynchus* share many features of their biology and ecology, this book contains numerous references to *Oncorhynchus*, especially to the large amount of North American research on environmental requirements. Most of the general conclusions are relevant to both genera, even though set in a primarily European context.

Since the Second World War a number of books on the genus *Salmo* has been published. The first major contributions were by Jones (1959) on salmon and by Frost & Brown (1967) on trout, followed by Mills (1971) on salmon and trout (but also including information on other related genera). These three volumes became standard works when published but, after a quarter of a century, they are now distinctly dated. More recently, there have been four books on single species. Mills (1989) and Shearer (1992) deal with the natural history, biology and management of Atlantic salmon. Mills (1989) includes lengthy and detailed sections on human impacts and possible means of mitigation. Baglinière & Maisse (1991) is a symposium volume on the biology and ecology of trout, and Elliott (1994) gives a detailed analysis of the ecology of the trout, though with its main emphasis on the population of one rather small stream. Up-to-date information on salmon in the sea is included in Mills (1993). The present book is intended for a comparatively wide readership. The author hopes that the later chapters will be of interest to fisheries scientists and ecologists. The themes of environmental requirements, conservation and restoration are, however, likely to be of interest to a much wider group that includes biologists, sedimentologists and engineers who are concerned with stream and river restoration. Restoration is, necessarily, an inter-disciplinary activity and it is important for each of the participants to have some grasp of the other branches of expertise that are deployed. The book may also be of interest to some anglers and natural historians. In an attempt to meet the needs of those readers who are not specialists in fisheries science, the following provisions have been made:

(1) Efforts have been made to minimize and define fisheries jargon throughout the book.

(2)   A rather selective glossary has been provided to help with technical terms, though most are defined within the main text.

(3)   Chapters 1 and 2 give a brief general introduction to the biology and importance of trout and salmon.

(4)   Chapter 3 is an introduction to the major environmental variables that are mentioned in later chapters.

(5)   In order to give quantitative expression to the interrelationship between different variables it is necessary to use mathematical terms and equations. Where equations are given, the terms used in them have been defined and their units stated. For those readers for whom equations are a total 'turn off' there are, in most instances, references to appropriate figures that summarize the relationships in graphical form.

Following the three introductory chapters, the book brings together (Chapter 4) current information on the ecology of salmon and trout, with particular emphasis on the definition and, where possible, quantification of their environmental requirements and limitations. This is the essential scientific basis for any attempt to conserve or restore the habitat and/or environment of trout and salmon. As our knowledge of this subject is far from complete, the present state of knowledge has been critically appraised and attention drawn to gaps and contradictions in our understanding. Chapter 5 examines the impacts of human activities on trout and salmon and Chapter 6 addresses various issues relevant to conservation and restoration. These two chapters are not a 'handbook' of detailed methodologies and techniques for stream restoration. Instead they attempt to consider broad principles and to draw attention to apparent shortcomings in many recent 'restoration' projects. Chapter 7 is a brief look towards the future.

In writing a book of this type the author, inevitably, faces a conflict between a desire to give full coverage to the world situation and a desire to deal in more detail with issues that are closer to home and of which he has more direct knowledge. This book is written primarily within a UK context. This is particularly true of Chapters 3 and 5. Nevertheless, most of the underlying principles are of general application.

The material in this book is as up-to-date as is practicable but, as there is an exponential increase in the scientific literature on this subject, some parts of it may be dated – even by the time of publication. The literature cited has been selected for illustrative purposes but every effort has been made to reflect the consensus of research findings on each topic and to indicate any divergences of view that are apparent. Most of the relevant major review papers have been mentioned and it is hoped that the works cited will be sufficient to enable the reader to gain access to the huge volume of salmonid literature that exists.

The farming of trout and salmon is now a major industry in parts of Norway, Scotland, Ireland and other countries but is covered only in so far as it impinges upon wild populations.

Acknowledgements

Dr A. Gustard provided information on base flow indices. Drs S. Welton and M. Ladle and Messrs. W.R.C. Beaumont and B. Dear gave information on chalk stream chemistry and access to unpublished data on gravel composition. Drs J. Irvine and M. Nagata contributed to Chapter 2 and Tables 1.1 and 2.2. Dr J. Adamson (NERC Institute of Terrestrial Ecology), the Dales area of the Environment Agency and the NERC Institute of Freshwater Ecology authorised the use of some of the data in Tables 3.5 and 3.7. Mr M. Brown and Mr S.B. Grave provided information on forest cover. Information on abstraction problems was provided by Mr A. Streven and Mr J. Bass. FAO fishery statistics were provided by Ms A. Crispoldi.

Permission to reproduce Figure 1.2 was given by the Freshwater Biological Association and Dr P.S. Maitland. Figures 2.2, 2.3, 5.3 and 5.4 were taken from photographs by Prof. P.A. Carling, and Figures 2.4 and 2.5 from photographs by Mr T. Furnass; these six figures are reproduced by permission of the Freshwater Biological Association. Figures 5.5 and 5.10 are from photographs by Prof. M.D. Newson, Figures 5.16 and 5.17 from Environment Agency photographs by courtesy of Miss A. Sansom; Figure 5.11 was taken by Mr and Mrs P. Hill.

Professor P.A. Carling, Prof. J. Hilton, Mr E.D. Le Cren, Dr R.H.K. Mann, Prof. M. Newson and Miss A. Sansom were all kind enough to read and criticize all or parts of the draft; Dr R.H.K. Mann very kindly read the proofs.

The artwork was made by Mr Jeremy Bray who also scanned the photographs. The word processing was done by Mrs D.C. Crisp.

The author is deeply indebted to all of these people for their help and also to very many others who have, perhaps unwittingly, contributed ideas during a multiplicity of conversations and discussions over many years.

Chapter 1
Taxonomy, Distribution and Importance

(pink salmon and red herrings)

Summary

The family Salmonidae contains the genera *Thymallus, Brachymystax, Hucho, Salvelinus, Salmo* and *Oncorhynchus*. This book is concerned chiefly with salmon and trout of the genera *Salmo* and *Oncorhynchus*, especially during the freshwater parts of their life cycle. The genus *Salmo* has two species and is indigenous to the North Atlantic area. The genus *Oncorhynchus* contains trout and salmon that are indigenous to the northern Pacific area. Attempts to introduce them to other areas have had variable success. Salmon and trout are important as indicators of river quality, as providers of sport and as an economic asset. The genus *Oncorhynchus* is economically more valuable than the genus *Salmo* by two orders of magnitude.

Before examining the life cycle and ecology of trout and salmon it is important to set this group of fishes within a wider context. In particular, we should consider, in outline, their taxonomy, distribution and importance. Taxonomy is the science of describing, classifying and identifying organisms and we will look briefly at the family Salmonidae, including the salmon, trout and various close relatives, and also at the main features used to distinguish between the two members of the genus *Salmo,* namely the Atlantic salmon (*Salmo salar* L.) and the trout (*S. trutta* L.). The natural geographical distributions of the members of the genera *Salmo* and *Oncorhynchus* (Pacific salmon) will be described and attempts to transplant these various species will also be touched upon. Finally, the social and economic importance of salmon and trout will be mentioned.

1.1   Taxonomy

The taxonomy of the Salmonidae has been modified frequently and, as Elliott (1994) noted, it will probably soon be reorganised again. The family Salmonidae contains the graylings (Thymallinae) as well as the charr, trout and salmon (Salmoninae). The Salmoninae comprise five genera.

1

Brachymystax (lenok) has only one species. It is a primitive form that occurs in Siberia (Frost & Brown, 1967) and is confined to fresh water.

Hucho has four species. The best known is *Hucho hucho* (L.), the Eurasian huchen or Danube salmon. This species occurs in parts of the Danube system and in some large Asian rivers. It is a large predatory species that spawns in shallows. As these fish grow larger they live in large deep pools. It is a threatened species (Holcik, 1990). Two other species are confined to large rivers in China and Korea. The fourth species lives in the Sea of Japan and visits some Japanese rivers to spawn (Holcik *et al.*, 1988).

Salvelinus (charr) has six species, of which the brook trout (*S. fontinalis* (Mitchill)) and the Arctic charr (*S. alpinus* (L.)) are the best known. The brook trout is native in streams in eastern North America where it fills a similar niche to that ocupied by *Salmo trutta* in Europe. It has been introduced to Europe from time to time but has established itself as a breeding species in very few sites. The Arctic charr has a circumpolar distribution. In the southern part of its range it usually occurs as a landlocked form in deep lakes. It is there regarded as a glacial relict and is usually referred to as the 'Alpine charr'. Further north the species is a river dweller, which spends part of its life at sea.

Salmo (salmon and trout) contains the two common European species of Salmoninae, namely the Atlantic salmon (*S. salar* L.) and the trout (*S. trutta* L.).

Oncorhynchus (salmon and trout) contains two trout species of western North America, one Japanese trout and six species of Pacific salmon (Table 1.1). It contains several additional trout forms from southwestern North America.

Table 1.1 Members of the genus *Salmo* and nine members of the genus *Oncorhynchus*.

Genus	Species	Common names
Salmo	*trutta* L.	brown trout, sea trout, trout
	salar L.	Atlantic salmon
Oncorhynchus	*clarki* (Richardson)	cutthroat trout
	mykiss (Walbaum)	rainbow trout, steelhead
	masou Brevoort	masu salmon, cherry salmon, sakuramasu, yamane, yamabe
	rhodurus (Gunther)	amago
	kisutch (Walbaum)	coho salmon, silver salmon
	tshawytscha (Walbaum)	spring salmon, chinook salmon, Quinnat salmon, king salmon
	keta (Walbaum)	chum salmon, dog salmon
	gorbuscha (Walbaum)	pink salmon, humpback salmon
	nerka (Walbaum)	sockeye salmon, red salmon, kokanee

Stearley (1992) recognises 13 extant species in the genus, though the specific status of some of the trout forms is dubious (McPhail, 1997).

Attention in this book will be given largely to trout and salmon of the genera *Salmo* and *Oncorhynchus*, with particular emphasis on the former. Consideration will be given mainly to the freshwater phases of the life cycle.

Criteria used to distinguish between the parr of the two species of *Salmo* are listed by Jones (1959) and Frost & Brown (1967), whilst Maitland's (1972) key also includes smolts and adults. The main features are indicated in Fig. 1.1, the necessary diagnostic information is included in Table 1.2. Many of the features are illustrated in Fig. 1.2. It is important to note that identifications should not be made on the basis of any one feature in isolation. Some individuals may appear to be intermediate between the two species in terms of some features. For example, a fish with ten dorsal fin rays and thirteen scales between the adipose fin and the lateral line (Table 1.2) could be either a trout or a salmon and other features must then be used to make a final determination. It is also worth bearing in mind that some specimens may, indeed, be intermediate because hybrids between the two *Salmo* species do occur in nature and this will be considered later.

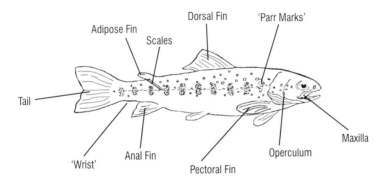

Fig. 1.1 Diagram to indicate some of the features used in specific identification of the genus *Salmo*, after Jones (1959) and Frost & Brown (1967).

1.2 Geographical distribution

The Atlantic salmon occurs on both sides of the Atlantic Ocean. It is found in rivers flowing to the northwest coast of mainland Europe from northern Portugal northwards to the White Sea, in the British Isles, Iceland and Greenland and on the eastern seaboard of North America from approximately 40°N northwards (Jones, 1959) to northern Quebec. In contrast, *Salmo trutta* is chiefly a European species. It occurs in Africa north of the Atlas mountains and in Europe (Frost & Brown, 1967; Elliott, 1994). It does not occur naturally in North America. *S. trutta* has both resident and anadromous (spending time at sea, returning to fresh water to spawn) forms. For convenience, in the rest of this book, the species will be

Table 1.2 Differences between the parr, smolts and adults of the two *Salmo* species. S = *Salmo salar*, T = *Salmo trutta*. Based chiefly on Maitland (1972).

	Parr		Smolt		Adult	
Feature	S	T	S	T	S	T
Parr marks	10–12	9–10	–	–	–	–
Spots on dorsal fin	few, faint	many, dark	–	–	–	–
Dorsal fin rays	10–12	8–10	10–12	8–10	10–12	8–10
Spots on operculum	<3	>3	<3	<3	–	–
Adipose fin	brown/grey	orange/red				
'Wrist' (caudal peduncle)	thin	thick	thin	thick	–	–
Tail	deep fork, pointed ends	shallow fork, rounded ends	deep fork	shallow fork	–	–
Pectoral fins	large	normal	large	normal	–	–
Maxilla	reaching to about middle of eye	reaching to between pupil and rear of eye	–	–	–	–
Scales between adipose fin and lateral line	–	–	10–13	13–16	10–13	13–16
Anal fin: when laid back does last ray extend about as far back as the first ray?	–	–	–	–	yes	no
Head of vomer bone toothed and shaft well toothed with persistent (rather than deciduous) teeth?	–	–	–	–	no	yes

referred to as the 'brown trout' in the form that is resident in fresh water and as the 'sea trout' in its anadromous form, whilst the unqualified word 'trout' or the scientific name will refer collectively to both forms. The brown trout is found throughout the range of the species but the sea trout is confined to the northern part of the range, from the Bay of Biscay northwards and to rivers afferent to the Black and Caspian Seas.

The genus *Oncorhynchus* (Table 1.1) occurs naturally on the Pacific seaboard of North America from California northwards, on the west coast of the Pacific Ocean from Taiwan northwards, and also on the east coast of Siberia. *O. clarki*

(a)

(b)

(c)

(d)

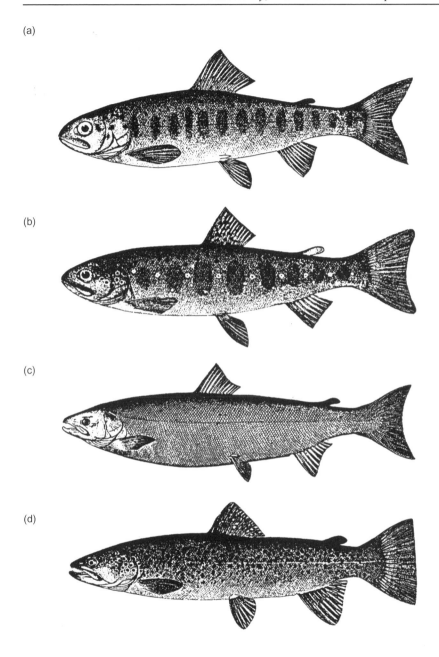

Fig. 1.2 *S. salar* parr (a) and adult (c), and *S. trutta* parr (b) and adult (d), from Maitland (1972).

and *O. mykiss* are trout whose natural distribution is confined to North America and, in common with *S. trutta*, they have both resident and anadromous forms. *O. masou* and *O. rhodurus* occur in Japan (the former also in East Siberia) and are considered by some authorities to be a single species. The remaining five species in Table 1.1 occur on both sides of the Pacific Ocean and are all 'salmon' in the sense that, with the exception of certain land-locked populations, they are all anadromous. Despite attempts to introduce the Atlantic salmon to various places outwith its natural distribution (*see*, for example, Arrowsmith & Pentelow (1965), on attempts in the Falkland Islands) there is rather little evidence of success. This may well reflect failure of the homing mechanism, especially after transfer to the southern hemisphere. In this context, it is interesting that Mills (1971) refers to a land-locked population of *S. salar* in South Island, New Zealand. These salmon have access to the sea, but do not use it. Instead, they descend the river only as far as Lake Te Anau and other smaller lakes but return to the rivers to spawn. Similarly, introductions of *O. nerka* to New Zealand produced only land-locked forms (Mills, 1989). Land-locked populations of *S. salar* occur in North America and Europe. The trout, *S. trutta*, now has a wide distribution in both northern and southern hemispheres (Table 1.3). This reflects the fact that in most trout populations some individuals always remain in fresh water and even the anadromous individuals are believed to stay much closer to their natal streams than do salmon. There is, thus, a lesser probability of extinction of the

Table 1.3 Recorded instances of successful introductions of the trout, *Salmo trutta* L., throughout the world, after a map by Elliott (1994) based on Arrowsmith & Pentelow (1965), Frost & Brown (1967), MacCrimmon & Marshall (1968), MacCrimmon *et al.*, (1970), Lesel *et al.*, (1971), Hardy (1972), Boeuf (1986) and Dumont & Mongeau (1989).

Locality	Years of introduction
Former USSR	1852–1939
Tasmania	1864
New Zealand	1867–1885
Sri Lanka	1883
USA	1883–1956
Canada	1887
Australia	1888
Kashmir	1889
Africa	1890–1932
Japan	1892
Pakistan	1903
South America	1904–1938
Madagascar	1926
Falkland Islands	1947
New Guinea	1949
Kerguelen Islands	1955
Nepal	1969

population as a result of changed geographical location disrupting the homing process. *O. tshawytscha* was successfully introduced to New Zealand in 1900 (Mills, 1971) and successful introductions of *O. nerka* and *O. kisutch* have been made to the U.S. state of Maine (Blair *et al.*, 1957). Landlocked forms of *O. nerka* occur and are known as 'kokanee'. Despite various attempts, however, no other Pacific salmon transplants appear to have been successful until large-scale Russian transfers of pink and chum salmon were made to Atlantic waters in the later 1950s, and coho salmon were introduced in parts of the Great Lakes system in the late 1960s. The success of these transplants appears to be somewhat dubious though there is some evidence of natural propagation (Mills, 1989). The rainbow trout (*O. mykiss*) is easy to propagate, grows rapidly and has, therefore, gained popularity amongst fish farmers. In Europe and elsewhere it is reared, often under rather intensive conditions, both for the table and for use in stocking 'put-and-take' sport fisheries. This species has been introduced to a number of countries, including Australia, India, Kenya, Malawi and South Africa, and has established breeding populations in most of them. In Europe it is widespread as a consequence of direct stocking and of escapes from rearing facilities. It may pose a threat to native salmonids and this will be considered later.

1.3   Importance

The importance of salmonids as a resource can be considered in terms of at least three criteria, though none of them lends itself to fully objective assessment.

The first criterion can be loosely described as 'moral, aesthetic and political' and relates to the fact that salmonids require water of good quality, and that their well-being in a particular river is one indicator of sound management and conservation of the fluvial resource. The presence or absence of salmonids is readily apparent to the public, many of whom derive pleasure from seeing large salmonids surmounting obstacles or smaller ones rising to surface foods, whilst others gain satisfaction simply from knowing that salmonids are present. This aspect of the value of salmonids is difficult to define clearly, let alone to evaluate.

The second criterion relates to the social value of salmonids as providers of sport fishing. This cannot be properly quantifed, even though its importance can be illustrated by quoting head counts and sums of money. The National Angling Survey (National Rivers Authority, 1994a) estimated that there were 841 000 game (salmonid) anglers in England and Wales. Estimates of annual direct expenditure by salmon and sea trout anglers in various parts of the British Isles (Table 1.4) should not be taken too seriously as absolute values, but they do indicate a relatively large measure of commitment to their sport by game anglers, especially as these sums exclude expenditure by those game anglers (probably a majority) who fish only for brown trout (and/or rainbow trout). We can, then, conclude that game angling is a major recreational asset.

Table 1.4 Estimates of annual expenditure by salmon and sea trout (not brown trout) anglers in various parts of the British Isles.

Source	Place	Expenditure (£ sterling × 10^6)
Scottish Tourist Board & the Highlands and Islands Development Board (1989)	Scotland	33.6
Department of Agriculture and Fisheries for Scotland (1982b)	Scotland	34.0
Whelan & Marsh (1988)	Eire	28.3
Radford *et al.* (1991)	England and Wales	16.4

The final criterion is a financial one. It is important to note that the precise values obtained by economic analysts depend heavily upon the methods of collecting and analysing the data and upon the assumptions underlying the calculations. Therefore, the results of such estimates cannot, necessarily, be compared validly between surveys and are best accepted simply as helpful orders of magnitude from which general conclusions can be drawn. If we indulge in sifting the fine details we will probably end up pursuing red herrings and lose sight of the broader implications. It is clear from Table 1.5 that the value of salmon and sea trout fishings in Great Britain is substantial and that, purely in terms of sale value, the recreational fisheries are more valuable than the commercial fisheries. The issue of whether or not a rod-caught salmon is of more value than a net-caught salmon to the national economy is much more complex (*see*, for example, Stansfeld, 1989) and will, for the sake of peace and brevity, be classed by this author as a further red herring! It is, however, important to recognize that both types of fishery can have considerable local importance in the creation of employment, especially as most of the game fishing rivers in the British Isles are in areas that are relatively disadvantaged economically. The issue of nets versus rods (and they are not, necessarily, mutually exclusive) requires careful analysis based on sound biological, economic and social statistics. It is also arguable that some ancient forms of net fishing (e.g. the haaf nets of SW

Table 1.5 Estimates of total nett economic value of recreational and commercial fisheries for salmon and sea trout based on 1988 returns, from Radford *et al.* (1991). Values in millions of pounds sterling. Note that this does not include brown trout fisheries.

	Recreational	Commercial	Totals
England and Wales	72.0	3.0–5.5	75.0–77.5
Scotland	255.0	6.0–11.0	261.0–266.0
Total	327.0	9.0–16.5	336.0–343.5

Scotland and NW England, and the coracle fisheries in Wales) should be preserved for heritage reasons. Game angling may contribute to the national balance of payments through the expenditure incurred by visiting anglers from other countries. In Great Britain this is probably negligible (0.3% of angler spending in England and Wales, 3.7% in Scotland), but in Eire approximately 45% of the estimated £28 million spent by anglers (Table 1.4) was by visitors to the country.

So far, the emphasis in this section has been upon the situation in the British Isles, though there is a notable absence of quantitative information on the size or value of the sport fishery for brown trout. Similarly, within a world context, the sea trout is somewhat neglected in the statistics because it is confined mainly to northern Europe and, in some countries, sea trout catches are included in the catch returns for Atlantic salmon. Annual catch statistics for Atlantic salmon are collected by the United Nations Food and Agriculture Organisation (FAO). The average annual catch in 1964 to 1968 was 13.2 thousand metric tonnes, with a value of about US $22 million. By the 1990s the total catch had fallen below five thousand metric tonnes (Fig. 1.3) and the distribution of the catch between different countries had also changed (Table 1.6) with smaller shares being taken by Canada, Denmark, Greenland and the UK, and larger shares by Finland, Iceland

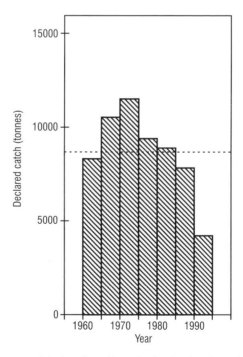

Fig. 1.3   Five-year averages of declared world catch of Atlantic salmon from 1960 to 1994. The broken line indicates the mean value over the period 1960 to 1994, inclusive, of 8675 tonnes. Derived from NASCO (1996).

Table 1.6 Declared weights of Atlantic salmon landed in various countries for the period 1987–1996, inclusive. The results are also expressed as percentages of the total. Note that catches by Estonia, Latvia, Lithuania and the Russian Federation in 1987 were combined as the USSR catch. The 1987 USSR catch has been divided between these four countries in the same proportions as their separate catches in the period 1988–1996. Comparable percentages for the period 1964–1968, inclusive, are also shown. Information from the FAO/FIDI Statistical Database.

Country	Annual mean catch (1987–1996) in metric tonnes (thousands)	Percentages of catch	Percentages for 1964–1968
Canada	0.54	5.7	17.7
Denmark	0.71	7.4	15.1
Faroe Islands	0.12	1.3	0.8
Finland	1.87	19.7	4.1
France	0.03	0.3	–
Germany	0.05	0.5	1.7
Greenland	0.28	2.9	8.3
Iceland	0.71	7.4	1.3
Ireland	0.90	9.4	10.1
Norway	1.55	16.2	13.1
Poland	0.29	3.0	1.5
Sweden	0.99	10.5	4.1
UK	0.72	7.5	15.4
Estonia	0.04	0.4 ⎫	
Latvia	0.34	3.6 ⎬	6.8
Lithuania	0.04	0.4 ⎪	
Russian Federation	0.36	3.8 ⎭	
Total	9.54	100.0	100.0

and Sweden. Since 1987, efforts have been made to estimate the size of the undeclared catch and the results suggest that approximately 28% of the total catch is undeclared (values for individual years vary from 24% to 32%).

Salmon have been an historically important source of protein in the human diet and, more recently, they have been a luxury food. During the last two decades there has been a rapid increase in the production of farmed salmon (Fig. 1.4) so that by 1995 the annual output of Atlantic salmon farms was approximately one hundred times the estimated catch of wild salmon (compare with Fig. 1.3). The main producers of farmed salmon are Norway (67.5% of the total between 1991 and 1995) with smaller industries in Russia, USA, UK, Eire, Canada, Iceland, Chile, Australia and the Faroe Islands. This massive production of farmed salmon has reduced the price of salmon to below that of many staple sea fish. It is important to note that the commercial fishery for Atlantic salmon is dwarfed by the commercial fishery for Pacific salmon. Ricker (1954a) estimated the annual catch of Atlantic salmon in Canada and the USA as 2.3 million kg, compared with 68.2 million kg of Pacific salmon in British Columbia and 454.5 million kg of

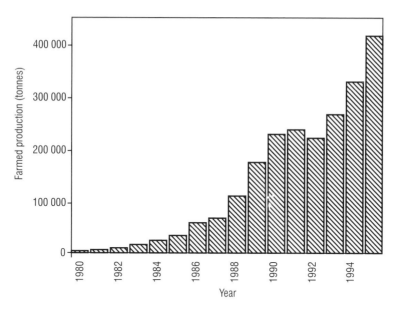

Fig. 1.4 Growth in the production of farmed salmon in the North Atlantic area between 1980 and 1995, after NASCO (1996).

Pacific salmon on the western side of the Pacific. Although now out of date, these values are still a useful indication of the relative importances of the two genera in a world context.

Chapter 2
Life Cycles and General Biology
(plastic fish!)

Summary

A basic life cycle is common to all trout and salmon. The female parent deposits eggs in the gravel. The eggs develop within the gravel and hatch to give alevins. The alevins remain in the gravel and subsist upon their yolk sacs. When the yolk is almost exhausted the alevins emerge from the gravel to become fry. The fry disperse from the oviposition site and become known as parr. They grow, become sexually mature and return to the area of the oviposition site to spawn. Within and between species there is a wide variety of variations upon this theme. These variations relate especially to the distance of dispersal of the juveniles with some staying close to the oviposition site and others going to sea, and to the duration of the different parts of the life cycle, especially the amount of time spent at sea. During free swimming life in fresh water, trout and salmon defend territories. At low initial densities, mortality/dispersal during the early months is proportionate (density-independent mortality/dispersal) but at higher densities the rate of mortality/dispersal increases with initial density (density-dependent mortality/dispersal) and thus has the effect of regulating population numbers. Trout and salmon are opportunist predators. They may feed both by foraging within their territories and by taking drifting prey.

This chapter indicates the wide variety of life cycle strategies that occur within and between species in the genera *Salmo* and *Oncorhynchus* and then considers aspects of behaviour and population dynamics, with special reference to the genus *Salmo*. Each of the two latter topics is worthy of a book of its own. Therefore, the present account is a general introduction but also contains references to more detailed works.

2.1 Life cycles

The most notable feature of the life cycles of trout and salmon is that a basic pattern is common to all species, but between and, even, within species, there is a wide range of variations upon this theme. This, together with the development of a range of regional terms for the different life stages, leads to some problems of definition. For present purposes we shall concentrate, initially, upon the two species of *Salmo* during their life in fresh water and then go on to consider their sea life. Finally we will examine the additional variations on the theme shown by some of the *Oncorhynchus* species. To describe the different life stages of the two *Salmo* species we will use the terminology summarized by Allan & Ritter (1977).

The freshwater part of the life cycle begins when the female parent selects a place where there is clean, flowing water and gravel of suitable size and composition. She excavates a 'pit' in the stream bed by means of a series of repetitive 'cutting' actions that involve turning on her side and making exaggerated swimming-like motions of her body. In this way, she creates suction that lifts gravel particles off the stream bottom. This gravel is displaced slightly downstream by the current. Thus a pit is formed in the streambed with a 'tail' of displaced gravel just downstream. From time to time she 'crouches' in the pit and lowers her anal fin. Apparently, this is to test the flow of water in the bottom of the pit. She may find conditions in the pit in some way unsatisfactory and will then abandon that site (leaving a 'false redd' or 'trial scrape') and seek another. When a pit has been made at a site that the female finds satisfactory, she deposits a batch of eggs in it; at the same time, the male (or males) releases sperm ('milt') over them. The female then cuts another pit immediately upstream of the first one and the gravel displaced from the second pit covers the eggs in the first. This process is repeated until one or more 'pockets' of eggs have been laid in a structure known as a 'redd' (Fig. 2.1). A female may produce one or more redds in a spawning season, but the most usual number is one per female. It is generally believed that the presence and behaviour of the male fish encourages the female to cut. Nevertheless, the presence of a male is not essential for completion of the process. A female sea trout has been seen to cut for several hours before finally being joined by a male (Crisp & Carling, 1989), and female Atlantic salmon in Canada have been known to construct redds and deposit eggs in the absence of any male (Myers & Hutchings, 1987). The above description is common to all species of stream spawning salmon and trout. More detailed accounts of the spawning behaviour of the *Salmo* species are given by Jones & King (1949, 1950) and Jones & Ball (1954). Although the processes are fairly straightforward, in practice the spawning of the *Salmo* species can be a protracted and rather chaotic procedure. The redd is usually attended by a dominant male who seeks to defend the redd site and female against other, usually smaller, males that seek to take part in the spawning. The coming and going of various males and the resultant threat posturing and jostling leads to frequent interruption and complication of the basic interaction between the female and the dominant male. Most of these encounters between males are resolved by means of threat and jostling, although Jones & Ball (1954) refer to males biting one another and to smaller fish being seized and shaken by larger males. This latter type of encounter has also been

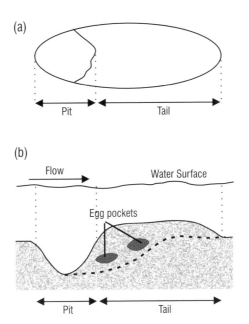

Fig. 2.1 Plan view (a) and longitudinal section (b) of a redd. The broken line marks the approximate limits of the disturbed gravel. After Crisp (1993a).

observed amongst Scandinavian sea trout by Bohlin (1975). At the completion of spawning most males usually move away from the redd and may seek to spawn with other females. In contrast, the female may remain for a while at the redd and may cut in a rather random and increasingly desultory manner around the edges of the final pit before moving away. The end product is a redd similar to that shown in Figure 2.1; the size of the redd is related to the size of the female. For most practical purposes a *Salmo* redd is approximately 3.5 times the length of the female fish, and its width 0.3 to 0.6 times her length (Crisp, 1996a). A more rigorous approach is that used by Crisp & Carling (1989) in which

$$\ln T = b \ln L + \ln a \tag{1}$$

where $\ln T$ is the natural logarithm of the length of the redd tail (cm), $\ln L$ is the natural logarithm of the length of the female fish (cm) and b and $\ln a$ are constants. For the genus *Salmo*, $b = 1.2 \pm 0.2$ (95% C.L.) and $\ln a = 0.45 \pm 0.38$ (95% C.L.). By use of this equation the length of the redd tail can be predicted from the length of the female fish. The tail length can then be used to predict other dimensions of the redd by use of the equations and constants given in Table 2.1. Such predictions have practical value. For example, when female spawners occur on gravel beds at high density, it can be useful to know the approximate areas required for redd construction by females of known sizes. High population

Table 2.1   Values of the constants ln a and b for substitution in the equation ln y = b ln *T* + ln a, where ln y is the natural logarithm of the named redd dimension (cm) and ln *T* is the natural logarithm of redd tail length (cm).

y	ln a ± 95% CL	b ± 95% CL
Pit length (cm)	0.46 ± 0.66	0.80 ± 0.13
Pit width (cm)	0.35 ± 0.71	0.80 ± 0.14
Tail width (cm)	0.43 ± 0.71	0.81 ± 0.14

densities on limited spawning areas can lead to later spawners damaging the redds of earlier spawners by 'overcutting' them.

The early part of the life cycle occurs within the gravel and the stages involved are termed 'intragravel' stages. The embryological aspects of early development are well-known because they can be studied readily in gravel-free conditions in hatcheries. We know much less, however, about the distribution, behaviour and survival of the intragravel stages in their natural habitat. One valuable step towards gaining more knowledge about the intragravel stages has been the development of methods of obtaining relatively undisturbed frozen cores of natural river gravel (Stocker & Williams, 1972). Their development for use in the UK (Carling & Reader, 1981) for collection of gravel samples for analysis (Fig. 2.2) led on to use in the study of the burial depth of salmonid eggs (Ottaway *et al.*, 1981; Crisp & Carling, 1989; Fig. 2.3). It is likely that freeze coring has further potential for use in studies of the distribution and movements of alevins within the gravel. The duration of the intragravel stages is determined largely by temperature. The first obvious change occurs when the eyes of the developing embryo become clearly visible through the shell of the egg (Figs 2.4(b) and (c)). The egg hatches (Fig. 2.4(d)) into an 'alevin' (Fig. 2.5(a)), which remains in the gravel and carries the residual egg yolk in the form of a yolk sac. As the alevin grows, it consumes the yolk and the yolk sac reduces in size (Fig. 2.5(b)). When the yolk sac is almost exhausted the alevin emerges from the gravel, fills its swim bladder with air to attain neutral buoyancy and begins to take external foods. This event is known as 'swim-up'. It marks the end of intragravel life and the young fish is then termed a 'fry' (Fig. 2.5(c)).

The fry soon disperse from the redd site, adopt feeding stations and establish and defend territories. They are then termed 'parr'. From this stage onwards the wide variety of life strategies shown by the two species of the genus *Salmo* becomes apparent.

Atlantic salmon remain in the river as parr for one or more years. The duration of this stage appears to be generally shorter in rivers to the south of the geographical range than in those to the north because growth rates are generally faster in the south. At the end of the parr stage the fish become 'smolts'; the age at which smolting occurs appears to be related, in part at least, to their size. Smolting usually occurs at lengths of 12.5–17.0 cm and is marked by body

Fig. 2.2 A freeze core of stream gravel being collected. Liquid carbon dioxide is injected into a standpipe driven into the gravel. This freezes the intragravel water to give a frozen core of relatively undisturbed gravel. Photograph by courtesy of Prof. Paul Carling.

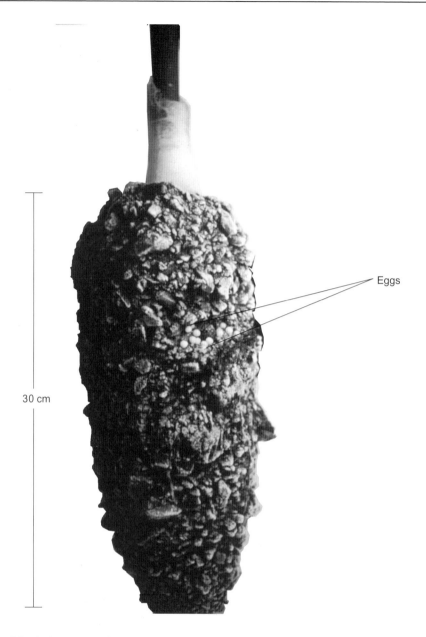

30 cm

Eggs

Fig. 2.3 A frozen gravel core containing a salmonid egg pocket. Photograph by Prof. Paul Carling.

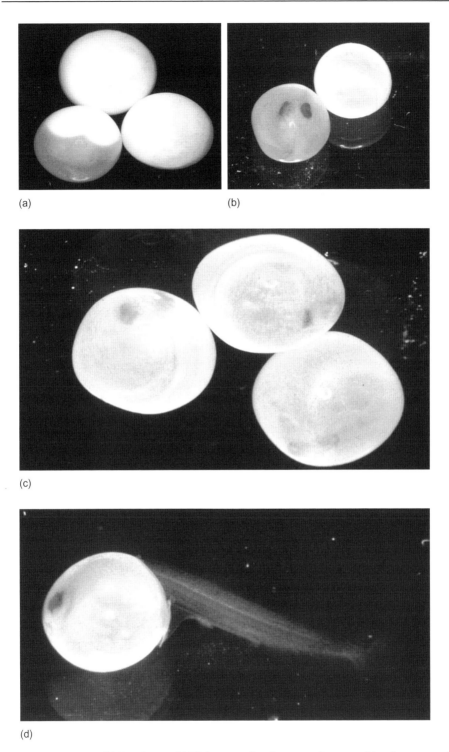

(a)

(b)

(c)

(d)

Fig. 2.4 Trout eggs. (a) Dead eggs. (b) Living eggs. One is at an early stage of development and the other is 'eyed'. (c) Eyed eggs with the embryos clearly visible. (d) An alevin in the process of hatching. Egg diameters are approx. 5 mm. Photographs by Trevor Furnass, ARPS.

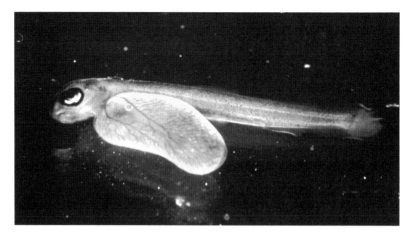

(a)

(b)

(c)

Fig. 2.5 Trout alevins and fry. (a) A recently hatched alevin. (b) An older alevin with partly absorbed yolk sac. (c) A fry shortly after emergence from the gravel. Alevin lengths are 15–20 mm. Photographs by Trevor Furnass, ARPS.

silvering and loss of parr marks (Fig. 1.1) and by darkening of the edges of the pectoral and caudal fins. The smolts move downstream to the sea. The fish may then remain at sea for one winter and return to fresh water to spawn as 'grilse' (or 'one-sea-winter fish'), or they may stay at sea for one or more additional winters to return to spawn as 'multi-sea-winter fish'. Very rarely, individuals may return as 'nought-sea-winter fish'. A proportion of male salmon parr becomes sexually mature in fresh water and remains there to attempt to participate in the spawning of those mature fish that return from the sea. These are described as 'precocious male parr'. After spawning a salmon is described as a 'kelt'. Most kelts die but some re-enter the sea and may return to fresh water to spawn on one or more subsequent occasions.

Terminology for the trout (*Salmo trutta*) is similar to that used for the salmon except that some individuals remain in fresh water all their lives, and that one-sea-winter adults are not usually called 'grilse'. The trout demonstrates an even wider spectrum of life strategy options than does the salmon. At one end of this continuum we have sea trout. These smolt, go to sea, and return to fresh water to spawn after varying numbers of years at sea. Some may return to fresh water to spawn during the same year that they went to sea. These are chiefly males and are best described as 'nought-sea-winter fish'. These nought-sea-winter fish have a wide variety of localised popular names including 'finnock', 'whitling' and 'her-ling'. The incidence of kelts returning to the sea and then reappearing in fresh water as spawners in later years is much higher in sea trout than in salmon. The percentage of fish surviving to spawn more than once varies between rivers and between years but is typically about 5% for salmon and 20–30% for sea trout. It is not clear if this is related to the fact that adult sea trout feed in fresh water and adult salmon do not. At the other end of the scale we have the resident brown trout that may spend all of its life within a few metres of the redd site in which it was born and which may, itself, spawn close to that point of origin. Between these two extremes are found 'slob trout' which migrate downstream only as far as the river estuary, and a range of patterns of migration within and between rivers and lakes within the freshwater system (Northcote, 1997; Jonsson, 1985). These different patterns can lead to large differences between individuals in growth and size at sexual maturity, spawning habitat requirements, survival and fecundity (Crisp, 1994). There is a marked tendency for female trout to move downstream more readily and/or for longer distances than do males. In streams that contain only sea trout, the sex ratio is close to 1:1 at all life stages. When a stream contains both sea trout and brown trout, however, there is usually a preponderance of females amongst descending smolts and returning adults, with a corresponding preponderance of males amongst resident fish (Pentelow *et al.*, 1933; Le Cren, 1985). Such observations are a powerful reason to argue that the anadromous habit in this species is not entirely determined genetically and, hence, that brown trout and sea trout do not 'breed true'. It has been suggested (Solomon, 1982; Jonsson, 1989) that they are components of a single breeding population within those rivers where both forms occur. Similar behavioural differences between the sexes have also been observed amongst brown trout returning from the main river to their natal stream to spawn (Crisp & Cubby, 1978) and also in a stream and reservoir system (Crisp *et al.*, 1990; Crisp, 1994).

A feature common to both species of *Salmo*, and most species of *Oncorhynchus*, is a general tendency to seek to return to their natal river and, furthermore, to their natal tributary, to spawn. This is as true of brown trout from a lake, reservoir or main river visiting tributaries to spawn as it is of salmon or sea trout returning from the sea. Nevertheless, a minority of fish do spawn in the 'wrong' stream, either because they cannot gain access to their natal stream or because of some malfunction of their homing mechanism. This minority may be very important in establishing populations in places that have only recently become accessible and/or suitable for use by salmonids and also for re-establishing the species in places where it has been extinguished by some catastrophic event (Quinn & Tallman, 1987; Thorpe, 1994). Imprinting with the 'smell' of the natal stream during early life is an important element in the homing mechanism.

There is accumulating evidence of genetic differences between the runs of fish into different rivers and even into different tributaries of the same river. Such differences would not persist unless homing were, generally, rather accurate.

As the two *Salmo* species often occur together in the same river and have similar spawning behaviour, some hybridisation might be expected. In the River Piddle, Dorset, hybrid smolts and parr have been found (Solomon & Child, 1978) and spawning by a male salmon and a female sea trout has been seen (Crisp & Carling, 1989). Serological studies (Payne *et al.*, 1972; Solomon & Child, 1978) on a total of 9166 juveniles assumed to be salmon from rivers in Eire and the UK indicated that about 0.3% were hybrids. Verspoor (1988) observed a rate of 0.9% in eastern Newfoundland (where the trout is present only as an introduction), compared with 0.3% in the UK and 0.07% in Sweden. We should, however, accept these values with some caution because the Eire and UK value, at least, is based on the examination of supposed salmon only. This could be misleading because in hatchery hybrids the first generation hybrids closely resemble *S. trutta* (Alabaster & Durbin, 1965). Perhaps the most striking feature of this matter is the fact that the rate of hybridisation appears so low. This could indicate that hybrid embryos may be less viable than pure bred embryos of either species but it is also likely to reflect a degree of spawning segregation between the two species. In general, salmon show less willingness to spawn in small tributaries than do trout; they also tend to spawn a little later in the year. Some authors lay most stress on spatial segregation (e.g. Le Cren, 1985), others on temporal segregation (e.g. Heggberget *et al.*, 1988).

The genus *Oncorhynchus* shows an even greater variety of life cycle patterns than does the genus *Salmo* (*see*, for example, Willson, 1997); these are summarized in Table 2.2. At one end of the scale is *O. rhodurus* which normally

Table 2.2 Summary of important life history characteristics for species of *Oncorhynchus*. Note that life histories are much more variable in transplanted populations. * Hatchery-reared fish usually go to sea at 1 year of age and spend one to three years at sea. † Most of the freshwater life is spent in lakes. Kokanee are land-locked forms.

Species	Form	Freshwater (years)	Sea (years)	Multiple spawning
O. clarki	coastal resident	0–3 entire life	1–3	+
*O. mykiss**	steelhead rainbow	2–5 entire life	1–3	+
O. masou	masu	1–2	1–2 (usually)	–
	yamane	entire life	–	+
O. rhodurus	amago	usually entire life (1–2)	<1	stream type only
O. tshawytscha	'stream'	1–2 (mainly)	0–6 (usually 2–4)	extremely rare
	'ocean'	<1	2–6	–
O. kisutch		0–2 (usually 1 or 2)	0.5–2.5 (usually 1.5)	–
O. nerka†	sockeye	0–4 (usually 1–2)	1–4	–
	kokanee	entire life	–	–
O. gorbuscha		0	1–3 (usually 2)	–
O. keta		0	1–5 (usually 2–4)	–

spends its entire life cycle in fresh water, though some individuals go to sea for less than a year. *O. clarki*, *O. mykiss* and *O. masou* resemble *S. trutta* in having both resident and anadromous forms. At the other extreme are the pink (*O. gorbuscha*) and chum (*O. keta*) salmon, which spawn in the lower parts of river systems and whose fry drift downstream to the sea soon after swim-up. The remaining species of *Oncorhynchus* show life cycles similar to that of *S. salar* except that most do not appear to mature as parr. The exception is *O. masou*, which shows a pronounced tendency to mature as parr (Nagata & Irvine, 1997) of both sexes. *O. nerka*, *O. kisutch* and *O. tshawytscha* have been known to produce mature parr, but only rarely (Clarke & Hirano, 1995; Willson, 1997) except for the fact that land-locked *O. nerka* (kokanee) do, of course, mature without going to sea. As *S. trutta*, *O. clarki* and *O. mykiss* have both resident and anadromous forms, some males and females will, obviously, become mature without going to sea. *O. clarki*, *O. mykiss*, *O. masou* and *O. rhodurus* may survive to spawn on more than one occasion. Such multiple spawning has been observed only extremely rarely in *O. tshawytscha* and has not been observed in sea-going *O. kisutch*, *O. nerka*, *O. gorbuscha* or *O. keta*.

Perhaps the most striking feature of the life cycles of trout and salmon is the huge variety of strategies that is possible and, indeed, is used. These variations range from that of those forms that are confined to fresh water, to that of the pink and chum salmon that use fresh water only for spawning and incubation and spend most of the rest of their lives in the sea. Within species there is also a very wide range of strategies. One of the most varied is seen in the trout (*S. trutta*) which shows a very wide variety of degrees of migratory behaviour, and also differences between the sexes in willingness to migrate. The anadromous form may spend varying numbers of years at sea before returning to fresh water to spawn and may survive to become a multiple spawner. Consider the trout population of a hypothetical river that contains both brown and sea trout. Suppose that most of the parr that will go to sea do so after spending two winters somewhere in the river system, though some may go after only one river winter and some may remain in the river for three winters. Suppose also that sea trout first become sexually mature and return to the river to spawn after spending anything from nought to four winters in the sea. Therefore, the population of sea trout spawning for the first time will contain fish showing up to fifteen different combinations of river and sea winters (Table 2.3). In addition to this, suppose that the brown trout mainly reach first sexual maturity (and spawn) after two river winters but that some (mainly males) do so after only one river winter and others (mainly females) do so after three river winters. We then have another three categories of first time spawners, giving a total of eighteen different combinations. At this point the plot thickens because members of each of these eighteen categories have the possibility of surviving beyond the first spawning to become multiple spawners and spawn on one or more subsequent occasions. This increases the original eighteen categories by a factor of two or more. There is probably no real river in which this pattern is typical. Nevertheless, this simple

Table 2.3 Possible combinations of numbers of river and sea winters passed by first-time spawners in the hypothetical population of sea trout described in the text. The combinations are shown in the body of the table in the format 'x.y' where x = number of river winters and y = number of sea winters.

	Number of river winters		
	1	2	3
Number of sea winters			
0	1.0	2.0	3.0
1	1.1	2.1	3.1
2	1.2	2.2	3.2
3	1.3	2.3	3.3
4	1.4	2.4	3.4

exercise, using assumptions that are well within the realms of possibility, illustrates the wide variety of life-history patterns that is available to and can be used by this species. Similarly, Willson (1997) noted that some species of *Oncorhynchus* may show numerous different age categories at first sexual maturity. These include *O. nerka* (22 categories), *O. tshawytscha* (16) and *O. mykiss* (18). In contrast, *O. gorbuscha* shows very little variability: most mature at age two and very small proportions at one or three years of age.

The evidence also suggests that, in populations using a variety of life cycle options, the relative proportions using each option may vary over time in a cyclical manner.

In summary, we are dealing with a group that shows immense plasticity of life cycle both within and between species and this may well be the key to the survival of this somewhat primitive group of fishes. As many of the underlying biological principles are common to both genera and most species, much of the remainder of this book will concentrate on the genus *Salmo*, with references to *Oncorhynchus* as appropriate.

2.2   Behaviour and social interaction

The homing of adults to their natal stream and subsequent spawning behaviour has already been described. The present section concentrates on the behaviour of free-swimming juveniles and resident adults in fresh water. Detailed reviews of this aspect, with special reference to Atlantic salmon, are given by Gibson (1988, 1993), and detailed discussion of territoriality in trout is given by Elliott (1994).

Sea trout fry have been seen emerging from the gravel by night. They rapidly dispersed downstream from the redd site and then took up territories (Moore & Scott, 1988). Salmon fry are also known to emerge at night. Marty & Beall (1989) noted two waves of dispersal amongst salmon, one soon after emergence and another 10 to 20 days later when territories were assumed and defended.

Solomon & Templeton (1976) suggested that, for resident trout, there is a general lifetime pattern in five phases. During the first summer (0 to 6 months) there are downstream movements from the redd sites to nursery areas. During the following winter (6 to 15 months) there are further downstream movements from nursery areas to areas of 'adult growth'. From 15 months to spawning there are limited movements of adults. At the time of spawning there are upstream movements to spawning areas and these are followed by downstream movements after spawning. They also identified two components in the population, a largely static one and a smaller group of very mobile fish. This outline pattern is a useful guide, but it is, clearly, a simplification in that it does not accommodate the more complex movement patterns that can be shown by trout that live in lake and river systems or that live in rivers divided by obstructions.

Although the fry observed by Moore & Scott (1988) all dispersed downstream, there is evidence that, either as fry or as parr, some 0-group trout (Elliott, 1986) and salmon (Egglishaw & Shackley, 1973) disperse upstream. There are conflicting reports on the distances of dispersal by salmon fry/parr in their early months. Beall *et al.* (1994) found that in SW France, by October, some parr had dispersed over 2000 m downstream and a substantial number had moved between 1000 and 1500 m. These quite long distances contrast with the values generally seen in the UK. Egglishaw and Shackley (1973) found that, by mid-August, salmon fry planted in the spring had dispersed up to 150 m upstream and up to 900 m downstream. Kennedy (1982) obtained similar results and noted that over 70% of the survivors were within the 100 m immediately downstream of the planting site, whilst Marty & Beall (1989) found that 50% were within 50 m of the planting site. Crisp (1995) planted salmon fry in May and found that by September the survivors had dispersed up to 50 m upstream and 950 m downstream but most were within the 20 m immediately downstream of the planting site and 20–25% were upstream. He suggested that between-years variation in dispersal distance might be related to summer rainfall values.

It has been claimed that dispersal of recently emerged trout fry is of little importance because those that fail to gain territories close to the redd site will be displaced downstream and die (Elliott, 1984a), but Solomon (1985) suggested that dispersal of viable fry might be an appreciable component of observed losses. In those streams where redds are close together and survival to swim-up is high, those fry that fail to establish territories are likely to be repeatedly repelled by territory holders and perish. In streams where redds are widely spaced and/or survival to swim-up is low, however, the downstream dispersal of viable fry/parr can be an important mechanism whereby the population maximises its use of available habitat (Crisp, 1993a).

The size of salmonid territories increases as the fish grow. Newly emerged salmon held territories of 0.02 to 0.03 m^2, whereas larger parr had territories > 1.0 m^2 (Kalleberg, 1958). Salmon of 10 cm length held territories of 0.2 to 5.0 m^2 (Symons & Heland, 1978). Grant & Kramer (1990) used the equation

$$\log_{10} S = 2.61 \log_{10} L - 2.83 \tag{2}$$

to describe the relationship between fish length (L, cm) and territory area (S, m^2) for a number of salmonid populations (including the two *Salmo* species). This relationship was further developed (Marschall & Crowder, 1995) on the assumption that space was a surrogate for some vital resource, probably food availability. Such equations cannot, however, be more than a general guide because there is a wealth of evidence (e.g. Kalleberg, 1958) to show that territory size is also influenced by such factors as the amount of available cover, the amount of visual isolation (which is further influenced by the effects of water velocity in causing fish to hold station high or low in the water column) and by the irregularity of the stream bed.

In addition to its influence upon territory size, fish size also influences the choice of territory site. In general, both salmon (MacCrimmon, 1954) and trout (Kennedy, 1982) move into deeper water as they grow and this is likely to lead to some partitioning of resources between different year classes within a species. There is evidence of interspecific and intraspecific interactions and these will be considered in Chapter 4.

It is noteworthy that territoriality does not occur at all life stages or in all circumstances. As smolts, both salmon and trout cease to be aggressive and tend to 'school'. It is also readily apparent that at times of drought, stream trout aggregate in residual pools and, apparently, are not highly aggressive. Under the crowded conditions of hatchery rearing tanks, territoriality is suppressed and there is evidence that hatchery residence may partially suppress the territoriality of salmon for some weeks after release, though no similar effect has been observed in trout (Crisp, 1996b).

The fact that, in general, juvenile salmonids occupy territories and defend them against rivals suggests that we have a clear-cut basis upon which to build a simple density-dependent model of population regulation (*see* 2.3 Population density, survival and growth). Unfortunately, there are several observations that indicate a greater complexity within the system than might be expected. We have already noted that territory size does not, necessarily, have an overriding influence on population density because territoriality may be reduced or suppressed. Symons (1971) used high densities of salmon parr (10 cm fish at 30 or more per m^2) in experimental channels. One group of salmon occupied territories and were described as 'territorially dominant', the others were described as 'sub-dominants' and were distributed peripherally between the territories of the dominants. Varying the food supply had little effect on the population density or territory size of the dominants but it did affect the numbers of sub-dominants. It is not clear if a similar situation could occur in a natural stream at lower population densities or to what extent this is simply another manifestation of the continued partial suppression of territoriality after release from a hatchery. More recent work (Huntingford & De Leaniz, 1997) indicates that behavioural traits, regarding territoriality and aggression, that give success at moderate densities when there is competition for localised food may differ from those that give success at low densities in more complex and unpredictable circumstances. Elliott (1994) estimated the total area held as territories by 0-group trout as a percentage of total stream area. The values were very low: 3–7% in April/May and 9–15% in July. He concluded that 'a large part of the stream is unsuitable habitat for young trout' and noted that 'suitable habitats cannot yet be defined'. Paradoxes of this type will only be resolved in the light of further more detailed studies of some of the complexities of salmonid behaviour, as described by such authors as Jenkins (1969), Metcalfe *et al.* (1989) and Elliott (1994).

2.3   Population density, survival and growth

Consideration of how salmonid numbers are regulated is best done by examining a type of mathematical model known as a 'stock-recruitment curve'. In its ideal form, this is a graphical plot of parent 'stock' (number of spawners) on the x (horizontal) axis against 'recruits' (number of spawners produced in the next generation) on the y (vertical) axis. In practice, the parent stock or cohort may be plotted as number of spawners or as number of eggs laid, and the number of recruits is often represented by the number of smolts or late 0-group fish produced. A number of different mathematical expressions have been proposed to fit these relationships and these have been listed and described in detailed mathematical terms by Elliott (1994). There is contention about the precise shape of the curve and, indeed, as to whether or not a single pattern of curve is appropriate for all populations. The two most common models are the dome-shaped curve of Ricker (1954b) (Fig. 2.6(a)) and the asymptotic model of Beverton & Holt (1957) (Fig. 2.6(b)). Solomon (1985) made a critical appraisal of the evidence for the domed curve. He concluded that the data set of Elson & Tuomi (1975) for salmon in the River Foyle (Northern Ireland) did not justify the fitting of a dome-shaped curve, and that the domed shape of the salmon data set of Gee *et al.* (1978) for parts of the Wye system could be explained by spatial redistribution at the pre-smolt stage. Elliott (1984a, b) produced a family of curves relating number of eggs m^{-2} (parent stock) and population densities of 0-group trout at various times of year (recruits) in Black Brows Beck (northern England). The plot for 0-group in May/June is clearly domed but as Solomon pointed out, 'few would agree that fry of this stage constitute recruitment' and he suggested that the domed shape, for this stage, may simply reflect an influence of population density upon the rate of density adjustment. Elliott's curves for 0-group at later dates appeared to be relatively flat-topped. Solomon concluded that the evidence for a domed curve was 'thin' and that it is necessary to ensure that the whole period of density-dependent mortality is covered by the observations and that fish movements are fully taken into account. He then went on to consider evidence for a flat-topped relationship and concluded that the examples he quoted (Ricker, 1954; Le Cren, 1973; Watt & Penney, 1980; Buck & Hay, 1984; Chadwick, 1985) did not prove that the relationship was not domed but did indicate a sizeable 'flat' portion to the curve. Since Solomon's review of 1985 the Black Brows Beck data have been augmented and curves that appear to be dome-shaped (though with a wide scatter of data points) have been added for 0-group in August/September and for spawning females (Elliott, 1994). Gardiner & Shackley (1991) have published data for salmon in the Shelligan Burn (Scotland) and fitted a domed curve, though this is based on only six rather scattered data points. A domed curve was fitted to 17 years' data from the River Bush (Northern Ireland) and an asymptotic curve to 19 years' data from the Altnahinch stream (Northern Ireland) by Kennedy & Crozier (1995).

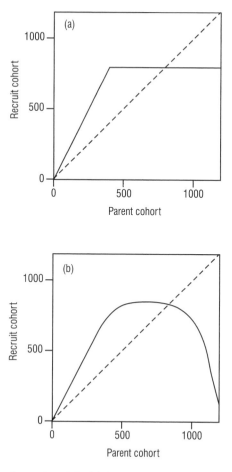

Fig. 2.6   Two 'stock recruitment' models. (a) Asymptotic model of Beverton & Holt (1957). (b) Dome-shaped curve of Ricker (1954b), modified from Solomon (1985). In the special case where the scales of both axes are the same and the same developmental stage is shown for both parent and recruit cohorts, points above and to the left of the broken (45°) line represent cohorts leading to recruit cohorts more numerous than themselves; points below and to the right represent parent cohorts that give recruit cohorts less numerous than themselves.

   Note that the curves have been simplified for illustrative purposes. In nature the two straight lines of Fig. 2.6a would merge as a smooth curve rather than meeting at an angle. Similarly, different authors give a number of variants upon the shape of the domed curve.

For most practical purposes, we can represent the situation adequately by means of the compromise scheme shown in Fig. 2.7. On the ascending portion of the graph (line a–b) the number of recruits is in constant ratio to the number of fish in the parent cohort or to the number of eggs laid by that cohort. This situation can be described as 'proportionate survival' or 'density-independent mortality'. As the size of the parent cohort increases further, this state of affairs is replaced by one in which the number of recruits is the same, despite increasing

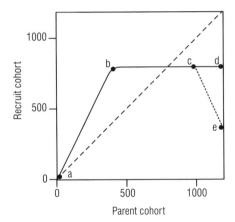

Fig. 2.7 A generalised 'stock-recruitment' model. a–b represents density-independent mortality (proportional survival), and b–d represents density-dependent mortality that brings pre-recruit numbers close to a fixed carrying capacity. The broken line is defined as in Fig. 2.6. The dotted line c–e represents the situation in which further increases in the parent cohort lead to reductions in the recruit numbers, though it is debatable whether or not this part of the model is often applicable to the genus *Salmo*.

Note that the curves have been simplified for illustrative purposes.

numbers in the parent cohort (line b–c, possibly line b–d). There are logical reasons to expect that at even higher parent densities the number of recruits will decrease as the parent stock increases. This would arise if, for example, the density of spawners on the spawning beds became so high that the spawners were overcutting each other's redds or were in some other way interfering with one another during spawning. Such mechanisms are represented by the line c–e. Such occurrences are probably rare in *Salmo* populations but may be more likely in some populations of *Oncorhynchus*. The line c–e is therefore included for completeness but will be disregarded in further discussion of Fig. 2.7.

Figure 2.7 is the key to understanding several important aspects of salmonid management. Consider, first, the flat-topped part of the graph (b–d). This situation occurs when there is overproduction of very young stages and density-dependent mortality operates. Death rate increases as fry numbers become higher and the population is thus reduced to fit the 'carrying capacity' of the habitat. It is sometimes argued that the density-dependent mechanism operates at a 'critical period' that corresponds approximately to the time when territories are assumed. We will accept this temporarily for the purposes of the present discussion. The recruit cohort number corresponding to the line b–d on Fig. 2.7 will reflect the carrying capacity at the 'critical period' but it will be less than the carrying capacity at that time because, between the critical period and the census of the recruit cohort, the fish will have increased in size and additional mortality (probably density-independent) will have taken place. In most published examples the data points show considerable scatter about the line b–d such that the highest observed values of recruitment on this part of the graph may be two or more times higher than the lowest observed values. This will, in part, reflect sampling errors in the estimation of fish numbers. It probably also includes a large element of year-upon-year variation in carrying capacity arising from between-years differences in variables such as temperature and rainfall. This wide scatter about the line may also help to explain the element of doubt as to the precise form of graph that is applicable to these sets of data. Despite the paradoxes described in the latter part of the previous section and the fact that carrying capacity may vary from year to year, it has value as a general working concept.

Documentation exists for several populations where the data points always or usually fall on the ascending limb (a–b) of the graph and the number of recruits varies widely from year to year. This implies that, in these populations, some form of density-independent mortality or restriction operates before the 'critical period' and with such intensity as to reduce the population density well below carrying capacity before the time when the 'critical period' is reached (in such circumstances the 'critical period' must be either negated or delayed).

At this point it is useful briefly to consider some aspects of a long term study on trout populations in streams around Cow Green Reservoir in the Pennine region of England (Crisp, 1993a) as this sheds further light on the regulation of 'edge of range' populations. Initial population was varied from year to year by varying the rate of stocking with swim-up fry. The numbers 'lost' between stocking and August could be apportioned between mortality and downstream dispersal by use of a downstream trap. In unmanipulated populations the August population density of 0-group trout varied greatly from year to year and was usually low (nought to two fish m^{-2}). In the stocked streams about 90% of the fry/parr were lost by August; the pattern of loss is summarised in Fig. 2.8. A hypothesis that explains these and other observations is given in Table 2.4. These studies showed that, in these streams, up to initial densities of ten swim-up fry m^{-2}, the rate of loss was proportionate and that downstream dispersal of viable parr was an important contributor to losses. It is not clear at what initial density value density-dependent mortality starts to operate in these populations but it is certainly above ten fry m^{-2}.

Rasmussen (1986) studied a Danish sea trout stream and suggested that density-dependent mechanisms related the numbers of I-group fish in November and the numbers of II-group fish in April to the numbers of 0-group fish 12 and 17 months earlier, respectively. Studies in the northern Pennines (Crisp, 1993a) and in a Welsh mountain stream (Crisp & Beaumont, 1995) have shown density-dependent relationships between 0-group trout numbers in July/August and the mean instantaneous rate of loss between then and the first three to five years of life (Fig. 2.9). Similar observations have recently been published by Elliott & Hurley (1998a). The studies by Rasmussen (1986) and Crisp & Beaumont (1995) refer to sea trout nursery streams, and Crisp (1993a) refers to streams that act as nursery streams for reservoir trout. In these three examples some of the fish 'lost'

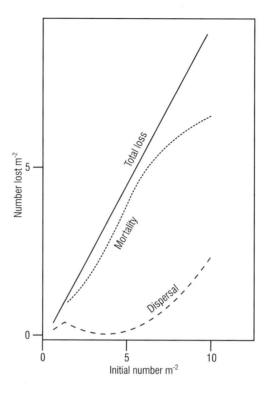

Fig. 2.8 Superimposed plots of total loss, mortality and downstream dispersal against initial density of swim-up fry for trout in a Pennine stream (northern England). After Crisp (1993b).

Table 2.4 Summary of a hypothesis to explain the dynamics of the patterns of mortality and dispersal shown in Fig. 2.8 (after Crisp, 1993b).

Initial density (Number m^{-2})	Behaviour
0–1.5	Parr are widely spaced, have few encounters with one another, show little territoriality and forage widely. The increased activity and wandering leads to a predominantly downstream shift and relatively high downstream dispersal rates.
3–5	Parr become increasingly territorial and are limited, in terms of food supply, to that produced in or drifting through their territories.
5–10	As population density increases, downstream dispersal occurs earlier and the dispersed fish are healthy and able to re-establish themselves elsewhere. The importance of dispersal relative to mortality, as a cause of loss, increases with initial population density.

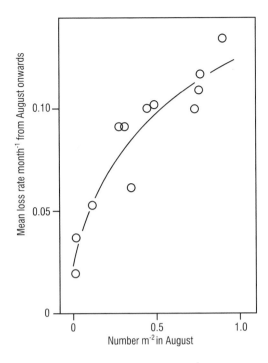

Fig. 2.9 Calculated mean instantaneous rate of loss month^{-1} of each trout cohort from August of its first year of life onwards, plotted against the mean number of 0-group fish m^{-2} in August, in a Pennine stream (northern England), after Crisp (1993a). Note that loss rate increases with August population density and the relationship is therefore density-dependent. The fitted curve accounts for over 90% of the variance of mean loss rate. Instantaneous rate of loss is a logarithmic rate. For further details, *see* Appendix A.

will have moved downstream to the sea or reservoir and may have returned later as large spawners. In the Elliott & Hurley (1998a) example, the population of trout was isolated above an impassable waterfall and any fish lost by downstream dispersal would certainly have made no further contribution to the upstream population. This subject is worthy of further study because such findings imply that the timing of the critical period can be very variable and may occur at any point in the life cycle at which the number/biomass of the survivors first exceeds the carrying capacity of the habitat. In this context, carrying capacity could refer to different limitations at different life stages, for example food and/or space for fry and juveniles and spawning sites for adults.

There is evidence for both *Salmo* species of inverse relationships between population density and the growth of 0-group fish. The weight attained by salmon in their first autumn in the Shelligan Burn (Scotland) could be related to September population density and water temperature (Egglishaw & Shackley, 1985; Gardiner & Shackley, 1991). Similarly, Backiel & Le Cren (1978) made experimental stockings of trout in Black Brows Beck (northern England) and showed an inverse relationship between growth rate and initial density, but only at initial stocking rates of < 20 swim-up fry m^{-2}. In contrast, Elliott (1994) could detect no relationship between egg density and the mean lengths or weights of trout in natural populations in the same small stream. The initial populations of eggs/alevins observed by Elliott were rarely less than $20\,\text{m}^{-2}$ and never less than $10\,\text{m}^{-2}$ and the information from Backiel & Le Cren suggests that it would, therefore, be difficult to detect such a relationship within Elliott's data. Data from Pennine streams (Fig. 2.10) show an inverse effect of population density on growth rate but this is only readily seen in these streams at August densities of less than 0.5 parr m^{-2} (i.e. about five swim-up fry m^{-2}). This initial density corresponds approximately to the value at which the effects of increasing territoriality are likely to become apparent (Table 2.4). The growth of young salmon and trout is likely to be influenced primarily by food supply and temperature, with population density as an ancillary. This explains why 90% of the variance of salmon growth could be explained in terms of population density and temperature (Gardiner & Shackley, 1991), whereas the use of population density alone accounted for only 40% of the variance of trout growth (Crisp, 1993a) and temperature alone can be used to predict growth rates that are between 60 and 90% of the observed values (Edwards *et al.*, 1979).

2.4   Feeding

In nature, trout and salmon are predators and, whilst feeding in fresh water, they take a wide variety of prey species, including aquatic animals and terrestrial casualties both at and below the water surface. In general, they feed

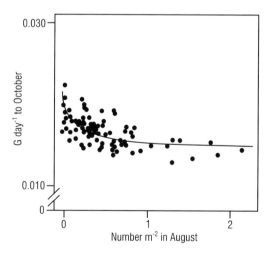

Fig. 2.10   Plot of mean instantaneous rate of growth in weight day^{-1} (G) from August to October against population density of 0-group trout in August, in a Pennine stream (northern England), after Crisp (1993a). The fitted curve accounts for 40% of the variance of mean growth rate. Instantaneous rate of growth is a logarithmic rate. For further details, *see* Appendix B.

opportunistically, though individuals and local groups may appear, at times, to specialise on particular foods. They may feed either by active foraging or by intercepting material that drifts close to a selected feeding station. Young trout and salmon take most prey animals up to the size with which they can cope physically. At the commencement of their free-swimming life they feed on very small animals, particularly micro-crustacea and very small larvae of Chironomidae and Ephemeroptera. As they grow, both the size and the variety of prey items increase. Eventually, a given individual may become large enough to feed on large prey items, including vertebrates such as fishes, amphibians and small terrestrial mammals. This may signal a change to a more rapid growth rate. In winter, juveniles may switch to night feeding and this will result in less efficient feeding but improved protection from predators (Metcalfe & Thorpe, 1992).

The diet of sea trout at sea includes a variety of organisms. The lesser sand eel (*Ammodytes tobianus* L.) is a major component, together with other fishes such as herring (*Clupea harengus* L.) and sprat (*Sprattus sprattus* (L.)) and various crustaceans and polychaete worms (Pemberton, 1976b; Fahy, 1983, 1985). Salmon take similar foods (especially sand eels and clupeioid fishes) whilst at sea. Growth of both species in the marine environment is rapid and exceeds the values predicted for the freshwater phase (*see* 4.2 Juveniles and smolts).

Adult sea trout feed in fresh water (Elliott, 1997) but adult salmon do not. Therefore, in some very long river systems, salmon (*Salmo* and *Oncorhynchus*) may face problems in reaching the spawning areas before they exhaust their body stores of energy. Such problems are, however, unlikely in rivers of more modest length such as all of those in the UK.

Chapter 3
Water Quantity and Quality and the Quality of Gravel
(swimming in murky waters)

Summary
Trout and salmon are affected by the quantity and quality of water flowing in a stream and the composition and form of the bed. The volume of water passing a point in unit time varies seasonally and also shows irregular short-term fluctuations in response to individual rainfall events. The size and rapidity of the short-term fluctuations varies between different types of river. Changes in flow are reflected in changes in water velocity, water depth and stream width. Natural waters contain a variety of materials in solution and suspension. Salmonid fishes are particularly sensitive to low concentrations of dissolved oxygen, low values of pH, high concentrations of suspended solids and the presence of various toxic materials. The water temperature in natural streams shows a seasonal cycle upon which are superimposed relatively regular daily fluctuations and irregular fluctuations arising from variations in sunshine and shade. Water temperatures can be modified by various human activities. Temperature influences many aspects of the biology of salmon and trout. The composition of stream bed gravel is of crucial importance to the spawning and incubation of trout and salmon. There is a variety of ways of characterizing gravels. Stream/river beds are complex, dynamic systems.

The quantity and quality of the water flowing in a water course and the composition of the bed material are the basic elements in the habitats of stream fish. As a further complication, these three factors, as will be seen, interact with one another. Some understanding of these matters is fundamental to making realistic assessments of the habitat requirements of fish.

To set the general comments in a realistic context, examples from real rivers and streams in the UK are given. There are, in fact, relatively few of the UK rivers for which there are good, long-term data on temperature, discharge and chemistry at a variety of points within the system. Fortunately, such data (together with information on gravel composition) are available for two examples that are close to the UK extremes. The chalk streams of southern England are in a mild climate, mainly spring-fed, with high concentrations of dissolved chemicals and with very equable temperature and flow regimes. In contrast, the River Tees is taken as an example of a northern river subject to sudden spates, with lower concentrations of dissolved chemicals and, in its headwaters, a relatively cold climate. It is, therefore, reasonably typical of UK streams rising in montane areas. Chemically the upper River Tees has what can be broadly termed 'hard' water during low flows and 'soft' water during high flows and is, therefore, not particularly extreme in chemical terms by UK standards. To fill this gap, chemical data are given from Afon Gwy, a small headstream in Wales that has very low concentrations of dissolved chemicals and very 'soft' water.

It is important to note that whereas the use of these UK examples suffices to illustrate the essential points, all the rivers of the UK are small (Ward, 1981) and are subject to a relatively equable climate in comparison to larger continental rivers.

3.1   River flow

'Flow' in a river or stream is measured as the volume of water passing a point in unit time and, to distinguish it from water velocity, is referred to as 'discharge'. A number of different units are and have been used to express discharge values. The most convenient for biological purposes are litres per second ($l\,s^{-1}$) for small quantities and cubic metres per second (cumecs or $m^3\,s^{-1}$) for larger quantities. These and three other units that are commonly encountered are summarized in Table 3.1, together with appropriate conversion factors. To set these units in perspective we can compare the maximum discharge of the author's cold bath tap and the mean discharge (between 1953 and 1990) of the River Spey which is the

Table 3.1   (a) Some units of discharge. (b) Relevant conversion factors.

(a) Units	
Millions of litres (megalitres) per day	$= Ml\,day^{-1}$
Millions of gallons per day	$= m.g.d.$
Cubic metres per second (cumec)	$= m^3\,s^{-1}$
Cubic feet per second (cusec)	$= ft^3\,s^{-1}$
(b) Conversions	
$1\,m^3\,s^{-1}$	$= 35.31\,ft^3\,s^{-1}$
$1\,m^3\,s^{-1}$	$= 19.01\,m.g.d$
$1\,m^3\,s^{-1}$	$= 86.31\,Ml\,day^{-1}$
$1\,m^3\,s^{-1}$	$= 1000\,l\,s^{-1}$

third largest river in Scotland and whose hydrology was described by Goody (1988). The bath tap delivers about 0.3 l s^{-1} (i.e. 0.0003 m^3 s^{-1} or 0.026 Ml day^{-1}), whereas the mean discharge of the River Spey is 64.5 m^3 s^{-1} (i.e. 5567 Ml day^{-1} or 64 500 l s^{-1}); this is similar to the mean discharge of the River Thames at Teddington (Butcher *et al.*, 1937).

An important feature of river discharge is that it has regular seasonal fluctuations. In temperate regions the flows of most natural rivers reflect seasonal rainfall patterns and have predominantly low discharges in summer and high discharges in winter. There are exceptions however. For example, rivers in relatively cold climates, where most winter precipitation is as snow and winter temperatures are generally below freezing point, may have low winter discharges followed by spates in the spring that are caused by rapid snowmelt. This situation is not common in the UK because, with the possible exception of high altitude streams in the Scottish Highlands, snowfall in the British uplands is unreliable from year to year in both depth and duration (Newson, 1981). It does, however, occur in some years in, for example, the upper River Tees (Crisp & Gledhill, 1970; Smith, 1971; Institute of Hydrology, 1975). Irregular short-term fluctuations, in response to individual rainfall events, are superimposed on the base flow and, in some rivers, these may be large and rapid. Rivers and streams fed mainly with water from limestone or chalk springs respond relatively slowly to rainfall, whereas upland streams that are fed largely by surface run-off can show very rapid responses and are termed 'flashy'. A good measure of the 'flashiness' of a stream or river is the 'Base Flow Index' (BFI) (Gustard *et al.*, 1992). This is a calculated index based on the relative contributions of 'base-flow' (water derived from groundwater and springs) and surface run-off to the total flow. A BFI of 1.0 implies no short-term fluctuations in discharge above the 'base-flow' and successively lower values indicate an increasing contribution by surface run-off and, hence, an increasing degree of flashiness. Table 3.2 shows values of BFI for several rivers and streams that are probably close to the observed extremes in the UK. The River Tees is one of the steepest (Institute of Hydrology, 1975; Newson, 1981) and most flashy (Ward, 1981; Gustard *et al.*, 1992) rivers in the UK and the

Table 3.2   Values of the base flow index (BFI) in parts of the River Tees system (from Gustard *et al.*, 1992), and in two chalk rivers.

River/Stream	Catchment area (ha)	BFI
R. Tees headstreams, upper Teesdale	–	0.15–0.34
R. Tees at Dent Bank, upper Teesdale	81 844	0.24
R. Tees tributaries downstream of Darlington	–	0.29–0.78
R. Tees at its tidal limit	193 000	0.39
R. Kennet	29 500	0.95
R. Test	104 000	0.95

very low values of BFI, particularly for the upper reaches and headstreams, reflect this. In 1937, when the upper Tees was less regulated by reservoirs than it now is, Butcher *et al.* stated that 'At times the rise is so rapid that the flood water passes down the river like a tidal wave, an effect that is known locally as the Tees Roll'. They also noted that 'the first waters of a flood are very turbid'. These points are illustrated well by Fig. 3.1, which shows the response of a Tees headstream to a thunderstorm in terms of discharge and eroded peat in suspension. In contrast, streams fed mainly by chalk springs have a large base flow component and respond relatively slowly to rainfall. This behaviour is apparent from the very high value of BFI for the Rivers Kennet and Test.

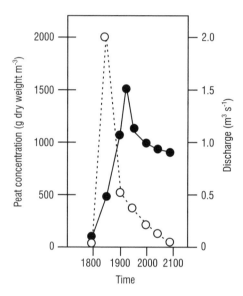

Fig. 3.1 Values of discharge and concentration of suspended peat particles at various times during a large spate, caused by a thunderstorm, in a small headstream of the River Tees. After Crisp & Robson (1982). Open circles = peat concentration, filled circles = discharge.

In fact, discharge can be considered in terms of three components, which may operate either separately or in combination to influence the value of a particular portion of watercourse as fish habitat. If we represent discharge as Q $(m^3 s^{-1})$ and its three components of mean stream depth, mean stream width and mean water velocity as D (m), W (m) and V (m s^{-1}) respectively (Fig. 3.2), then

$$Q = DWV \hspace{4cm} (3)$$

The relative changes in D, W and V with changing value of Q will depend mainly upon the shape of the stream cross-section and the gradient of the stream. In a stream with a steep-sided cross-section, increases in Q will give relatively small

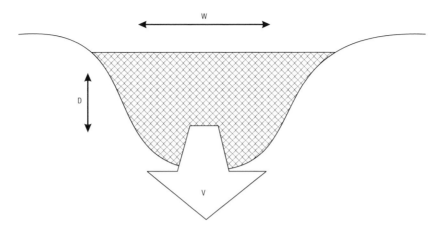

Fig. 3.2   Cross-section of a stream to show the three components of discharge (Q) such that Q = WDV, where W, D and V are mean values of water width, depth and velocity, respectively, for the cross-section.

increases in W and relatively large increases in D, whereas when the stream margins are gently sloped, there will be relatively large increases in W and relatively small increases in D. Similarly, in streams with a steep gradient, increase in V with increase in Q will be relatively large and changes in D and W will be relatively small.

So far we have considered the mean water velocity of a stream or river cross-section. Within that cross-section, however, there will be considerable spatial variation in water velocity. In general, water velocity will be slower close to the stream margins than in midstream. In addition, the water velocity close to the bed will be much less than that at the water surface, especially when the stream bed is rough. A good approximation to the mean velocity of the whole water column at a particular point is the measurement of the velocity at 0.6 of depth. Thus, if stream depth is 1.0 m, then the velocity at a point 0.6 m below the surface or 0.4 m above the bed is a good approximation to the mean value for the whole water column. Velocity at 0.6 depth is often used as a standard measurement for spatial or temporal comparisons of water velocity, though this may not hold when the bed contains large boulders.

As we will see in Chapter 4, water velocity, depth and fluctuations in flow are of major importance to all life stages of trout and salmon.

3.2   Water chemistry

Before considering water chemistry it is necessary to note the units in which various quantities are usually expressed. River water contains a wide variety of materials in solution and suspension. A selection of some of the most common inorganic materials found in solution is given in Table 3.3. The concentrations of these determinands are usually given in mg l^{-1} (where 1 mg = 0.001 g), but are also described as 'parts per million' (p.p.m.). Some determinands that are found only at very low concentrations are given in micrograms l^{-1} ($\mu g \, l^{-1}$), where 1 μg = 0.000001 g. Gases in solution may be quantified either as mg l^{-1} or as a percentage of the saturation concentration. A more complex concept is that of pH. The following brief summary is sufficient for present purposes:

(1) pH is a measure of the balance between hydrogen ions (H^+) and hydroxyl ions (OH^-) and is therefore a measure of the acidity or alkalinity of the water.
(2) A pH value of 7.0 represents neutrality (an exact balance between hydrogen and hydroxyl ions).
(3) Values above 7.0 indicate alkaline conditions, and values below 7.0 indicate acidity. The lower the pH, the more acidic is the solution and the higher the pH, the more alkaline.

Table 3.3 Concentrations (mg l^{-1}) of common ions/elements in rainfall in Co. Galway, Eire, and the English Lake District (from Gorham, 1957); and in river water from the Afon Gwy, Wales (from Patrick *et al.*, 1991 – three years' average), the River Tees (from Crisp *et al.*, 1974 – one autumn sample), and the River Frome (from Crisp & Gledhill, 1970 – two years' average). Approximate percentages of the rainfall concentrations that can be attributed to sea spray are indicated in parentheses.

| | Rainfall | | River water | | |
	Co. Galway	Lake District	Afon Gwy	River Tees	River Frome
Na^+ (sodium)	2.9 (89)	1.9 (95)	3.3	2.5	13.2
K^+ (potassium)	0.3 (31)	0.2 (37)	0.1	0.5	1.9
Ca^{++} (calcium)	1.6 (6)	0.3 (24)	0.7	18.3	84.0
Mg^{++} (magnesium)	0.5 (67)·	0.2 (100)	0.7	1.3	2.5
HCO_3^- (bicarbonate)	4.2 (<1)	0 (0)	–	–	–
Cl^- (chloride)	4.6 (100)	3.3 (100)	5.5	3.3	13.8
SO_4^{--} (sulphate)	2.3 (27)	3.2 (14)	3.5	8.0	30.5
NO_3^- N (nitrogen as nitrate)	0.05	<0.02	–	0.06	–
N (total oxidised nitrogen)	–	–	–	–	2.2
P (dissolved reactive phosphorus)	–	–	–	–	0.1
Si (dissolved reactive silicon)	–	–	–	–	3.4

(4)   pH is a logarithmic scale, therefore a change in pH of one unit represents a tenfold change in acidity/alkalinity.
(5)   In biological applications, pH values are not likely to fall outside the range 3.0 to 11.0.

Natural waters contain several major ions that are usually each present at concentrations of units, tens or hundreds of $mg\,l^{-1}$. These are sodium (Na^+), potassium (K^+), calcium (Ca^{++}), magnesium (Mg^{++}), sulphate (SO_4^{--}), chloride (Cl^-), carbonate/bicarbonate (CO_3^{--}/HCO_3^-) and nitrate (NO_3^-). The concentrations of these define the basic chemistry of river water. The salinity of the water usually refers to the concentration of common salt, sodium chloride (NaCl), in the water and this is high in sea water and low in fresh water. The 'softnesss' or 'hardness' of a water refers to the concentrations of calcium (Ca) and magnesium (Mg), especially the former. Calcium is very important in giving protection against the effects of certain pollutants.

The water in lakes and rivers falls as rain or snow and enters the watercourses either as surface run-off or after percolation through soil or rock formations. The precipitation takes chemicals into solution during its fall through the atmosphere and during its subsequent travel into the watercourse. Table 3.3 shows the concentrations of the major ions and some others in rainfall at two sites where industrial pollution of the atmosphere is likely to be minimal, though contamination of the rainfall by sea spray does occur. It is apparent that, if the sea spray component is deducted, the concentrations of these determinands in the rainfall are low. Gorham (1958a, b) showed that daily weather conditions had considerable influence on the chemical composition of rainfall and that industrial pollution of the atmosphere could be important in lowering the pH of rainwater to give 'acid rain'.

The concentrations of common ions and elements may vary considerably between rivers (Table 3.3) chiefly as a consequence of differences in geology and land use. The Afon Gwy (upper Wye) in mid-Wales has little or no input from calcareous rocks and the chemical concentrations differ little from those of rainfall. The River Frome (Dorset) is fed chiefly by chalk springs and derives its solutes mainly from chalk aquifers and various agricultural inputs. Chemical concentrations in the River Frome are high, relative to rainfall. The chemical composition of the water in chalk streams and rivers varies little with season or discharge (Casey, 1969; Casey & Newton, 1972). Concentrations in rivers of other types may, however, vary markedly with discharge. Most of the tributaries of the upper River Tees flow over a number of bands of limestone and the concentrations of various determinands vary with discharge, largely as a reflection of the amount of time that the water spends in contact with the limestone and the proportion of surface water in the discharge. In one Tees headstream there were complex relationships between fluctuations in concentration of potassium (0.25–$1.0\,mg\,l^{-1}$), phosphorus (0.01–$0.04\,mg\,l^{-1}$), nitrogen (0.1–$0.22\,mg\,l^{-1}$), calcium

$(1.0–8.5 \, \text{mg} \, \text{l}^{-1})$ and discharge (Crisp, 1966). Calcium concentrations in monthly samples from the upper River Tees in 1975 varied from $3.5 \, \text{mg} \, \text{l}^{-1}$ in high discharges to $37.2 \, \text{mg} \, \text{l}^{-1}$ in low discharges.

Natural stream and river waters may contain a variety of organic compounds in solution or colloidal suspension. They include, for example, humic materials originating from peat but others may arise from industrial or agricultural effluents. Our understanding of these is limited.

Fresh water contains various gases in solution. The chief ones are oxygen, nitrogen and carbon dioxide. Nitrogen and oxygen are present in the air in a ratio of five to one, but nitrogen is less soluble in water than is oxygen and the ratio in solution is three to one. The amount of any gas that can be held in solution by water depends upon the temperature (Fig. 3.3) and (of lesser importance in the present context) the atmospheric pressure. It is worth noting that whereas one litre of air contains about 200 ml of oxygen, one litre of water contains about 10 ml. As noted, dissolved oxygen concentration may also be expressed as a percentage of the saturation value. The amount of oxygen (as $\text{mg} \, \text{l}^{-1}$) that gives 100% saturation (ASV) is given by the equation

$$100\% \; \text{ASV} = 468 \, / \, 31.6 \; \text{T} \tag{4}$$

where T is temperature in °C (Montgomery *et al.*, 1964). It is worth noting that although the capacity of water to hold oxygen in solution decreases with rising

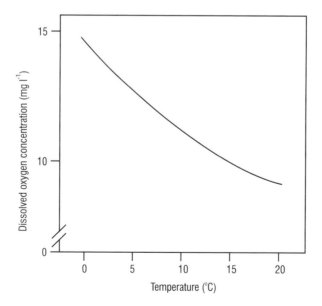

Fig. 3.3 Relationship between water temperature and the saturation concentration of oxygen at an air pressure of 760 mm of mercury.

temperature, the oxygen needs of fish increase with temperature. Dissolved oxygen concentrations may show diel (daily) fluctuations that reflect the photosynthetic activity of aquatic plants. The River Frome, Dorset, contains dense growths of submerged macrophytes (chiefly *Ranunculus* sp.), whereas the Maize Beck in upper Teesdale contains hardly any macrophytes but does have algae and occasional mosses. In the River Frome (Fig. 3.4) oxygen concentration is above saturation for most of a summer day but there is a distinct fluctuation from below 80% saturation in the early morning to over 160% saturation during the early evening. In Maize Beck there is a similar diel fluctuation with a peak around midday and a trough around midnight. However, the concentration fluctuations are very small (92 to 100% saturation) compared with those in the River Frome and values do not exceed 100%. Submerged plants consume carbon dioxide and produce oxygen as they photosynthesise during daylight. In the hours of darkness photosynthesis ceases and there is a net production of carbon dioxide. This explains the diel fluctuations in dissolved oxygen. It is also responsible in heavily vegetated streams (e.g. chalk streams) for daily fluctuations in pH. The carbon dioxide generated at night combines with water molecules to form carbonic acid (H_2CO_3); this ionises and thereby lowers the pH.

Various finely divided solids can occur in suspension. These may arise either from natural or anthropogenic sources. Concentrations may vary with discharge, often at a higher rate than the rate at which discharge fluctuates (e.g. *see* Fig. 3.1). There is usually interaction between suspended solids and the stream bed

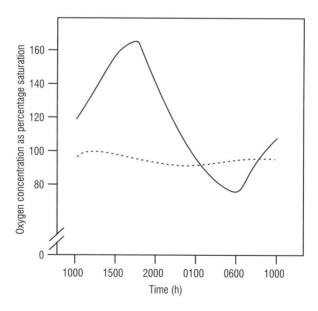

Fig. 3.4 Summer daily fluctuations in dissolved oxygen concentration in the River Frome, Dorset, England (solid line) (after Casey, 1969), and in Maize Beck, a tributary of the upper River Tees, northern England (broken line) (after Crisp, 1977).

material in the sense that these materials may be deposited in the stream bed during the recession of a spate and then be remobilised during a subsequent spate.

The influence of temperature, oxygen supply, inert suspended solids and pH upon various vital activities of salmonid fishes will be considered in more detail in later chapters. It is, however, convenient to consider now some of those materials that can be harmful or toxic to fish when present at too high a concentration. Inorganic materials (especially some metals) can be toxic to fish. Organic materials vary in their action. Some (e.g. cyanide) may be toxic. Other organic materials may, in the course of breakdown by microorganisms, rapidly consume oxygen and thus deplete the oxygen in solution and lead to the asphyxiation of aquatic animals. In considering the concentrations of various pollutants that can be harmful to fish, several complicating factors must be taken into account:

(1) The toxicity of a given pollutant may be modified by the pH and/or calcium concentration.
(2) The toxicity of 'cocktails' of pollutants is often, but not always, additive.
(3) The occurrence of death often depends upon some function of concentration and duration of exposure.
(4) Even sub-lethal concentrations of pollutants may be harmful in their effects upon such vital processses as the growth or reproductive efficiency of fishes.

The Council of the European Union (EU) has issued a directive on the quality of fresh waters and the European Inland Fisheries Advisory Commission (EIFAC) has produced guidelines. This and other information has been brought together by Alabaster & Lloyd (1982) and Solbé (1988). These works should be consulted for a detailed coverage. Some of the main limits and guidelines, with particular reference to salmonid fishes, are summarized in Table 3.4. Further information is given by Hellawell (1986).

3.3   Water temperature

In a typical stream or river, water temperature shows two main types of comparatively regular fluctuation. First, there is a seasonal cycle that has generally higher temperatures in summer and lower temperatures in winter. Second, there is a daily cycle with a minimum temperature during the night and a maximum during the day. Superimposed on the daily cycle, there may be various irregular fluctuations in response to changing amounts of clear and cloudy sky, to shading effects and to various precipitation events. Finally, the amplitude and/or mean value of the annual cycle varies between warm and cool years and may show longer-term fluctuations in response to global or regional changes in climate.

Table 3.4 Summary of information on limits and guidelines for toxins in fresh water, with special reference to salmonid fishes. Information mainly from Alabaster & Lloyd (1982) and Solbé (1988). † indicates an effect of water hardness or calcium concentration in modifying toxicity.

Determinands	Limits/guidelines	Notes
Finely divided inert solids	$25-80 \, mg \, l^{-1}$ acceptable, $< 25 \, mg \, l^{-1}$ preferable.	
pH	Harmful at < 5.0 and > 9.0, lethal at < 4.0 and $> 9.5-10.0$.	May influence toxicity of other poisons.
Ammonia	Toxicity is due to the unionized form (NH_3). The guideline value for NH_3 is $5 \, \mu g \, l^{-1}$ with a maximum permissible of $25 \, \mu g \, l^{-1}$.	Toxicity dependent on pH.
Chlorine	The toxic species is hypochlorous acid (HOCl) which is most abundant at low pH. Maximum permissible is $5 \, mg \, l^{-1}$ as HOCl.	Toxicity dependent on pH. HOCl favoured at low pH.
Cyanides	The toxic species is the unionized molecule (HCN). Lethal at about $4 \, \mu g \, l^{-1}$.	Toxicity dependent on pH. HCN favoured at low pH.
Nitrates	Toxic when reduced to nitrite (NO_2).	Contribute to eutrophication.
Nitrites	Highly toxic as the nitrite ion NO_2^-. EU directive level is $10 \, \mu g \, l^{-1}$ as NO_2 ($3 \, \mu g \, l^{-1}$ as NO_2-N). EIFAC proposes five levels ranging from $10 \, mg \, l^{-1}$ NO_2-N where only $1 \, mg \, l^{-1}$ of chloride present to $150 \, \mu g \, l^{-1}$ where $40 \, mg \, l^{-1}$ of chloride present.	
Phosphates	Not toxic but contribute to eutrophication. Hence EU limit is $0.2 \, mg \, l^{-1}$ as PO_4.	
Aluminium	Only toxic as the monomeric form Al^{+++} which occurs only in soft, low-pH waters and is toxic at concentrations of only a few μg.	A widespread and abundant element (e.g. in clay).
Cadmium	Lethal at low concentrations during long-term exposure. EU member states have agreed to eliminate pollution from cadmium and mercury.	
Chromium	Toxic as Cr_2O_7. Recommended mean concentration $< 25 \, \mu g \, l^{-1}$ and the 95 percentile should be $< 100 \, \mu g \, l^{-1}$.	

Cont.

Table 3.4 Cont.

Determinands	Limits/guidelines	Notes
Copper	Toxic as the cupric ion (Cu^{++}). EU sets four guideline values ranging from $5\,mg\,l^{-1}$ for soft waters ($<10\,mg\,l^{-1}$ as $CaCo_3$) to $112\,mg\,l^{-1}$ for hard waters ($>300\,mg\,l^{-1}$ as $CaCO_3$).	†
Iron	Only normally toxic in waters already dangerously acidic for fish.	
Lead	Lethal to rainbow trout at levels of about $1\,mg\,l^{-1}$ in water of medium softness ($50\,mg\,l^{-1}$ of $CaCO_3$) during 48-hour exposure.	†
Mercury	Exceptionally toxic in certain compounds. *See also* 'Cadmium'.	†
Nickel	Average concentration should be $<10\,\mu g\,l^{-1}$ and 95 percentile $<30\,\mu g\,l^{-1}$ in soft water ($20\,mg\,l^{-1}$ as $CaCO_3$) and the average and 95 percentile should be $<40\,mg\,l^{-1}$ and $<120\,mg\,l^{-1}$, respectively, in hard water ($320\,mg\,l^{-1}$ as $CaCO_3$).	†
Zinc	EU mandatory levels from 30–$500\,\mu g\,l^{-1}$ for waters with hardness values from 10–$500\,mg\,l^{-1}$ as $CaCO_3$.	†

Table 3.5 shows the annual temperature cycle in various parts of the River Tees system (northern England). As the water passes downstream from a small mountain headwater to the tidal limit, the annual mean temperature increases as do the monthly means. A predominantly surface-fed small stream in the south of England (Dockens Water, *see* Table 3.6) has an annual mean similar to that of the lower Tees, but rather smaller annual amplitude. In contrast, Bere Stream is a small chalk stream and the River Frome is a chalk river. Both are fed predominantly by chalk springs and the temperature of the water emerging from such springs is almost constant throughout the year. Two springs whose temperatures were examined in detail had annual mean temperatures of 11.0°C and 10.7°C and annual amplitudes of 1.5°C and 0.4°C, respectively. The two spring-fed watercourses have higher annual means and smaller annual amplitudes than do the stations on the River Tees, and these differences reflect partly a milder climate and partly the differences in type of source.

The size of the daily fluctuations in a stream can be characterised as monthly and annual means of the observed daily range (daily maximum–daily minimum). It is apparent (Table 3.7) that the daily range is larger in summer than in winter. It

Table 3.5 Monthly mean water temperatures on the River Tees, ranging from a small moorland tributary (Rough Sike) through various main river stations down to the tidal limit at Low Moor. The Rough Sike data are means for June 1993 to May 1996 and include the hot summer of 1995. The Tees Bridge data are means for 1973 + 1974 + 1975 from Crisp (1988a); the Dent Bank and Broken Scar data are means for 1962 + 1963 + 1964 from Smith (1968); and the Low Moor data are means for 1992 + 1993 + 1994. * indicates means that contain data for the very severe winter of 1962/3.

	Rough Sike	Tees Bridge	Dent Bank	Broken Scar	Low Moor
Altitude (m.O.D.)	565	527	226	37	8
Mean discharge ($m^3 s^{-1}$)	0.04	2.5	7.7	17.5	18.1
Jan.	0.7	1.9	1.3*	1.6*	2.6
Feb.	0.4	1.9	1.9*	2.2*	3.8
Mar.	0.8	3.1	3.4	4.2	5.9
Apr.	4.2	5.5	6.4	7.8	8.5
May	7.3	8.9	10.6	12.1	13.1
June	11.2	12.8	13.7	14.9	16.9
July	13.4	13.3	15.8	15.8	18.0
Aug.	12.4	14.1	14.8	14.9	16.0
Sept.	9.1	9.4	13.0	13.0	12.6
Oct.	6.4	5.3	7.6	9.4	8.4
Nov.	4.0	2.1	4.1	5.2	5.9
Dec.	1.3	1.6	1.7	2.2	3.9
Mean	5.9	6.7	7.9	8.6	9.6
Amplitude	13.0	12.5	14.5	14.2	15.4

is also larger in streams than in rivers and this probably reflects the larger thermal capacity of the latter. The timing of the daily cycle also varies with distance downstream. In a typical headstream the daily minimum temperature may occur around 0500 h to 0600 h, and the daily maximum around 1500 h to 1600 h. In a river (e.g. the River Tees at its tidal limit) the minimum may not occur until around 0900 h and the maximum around 2100 h.

The temperature of natural streams and rivers follows the air temperature and, apart from periods of prolonged sub-zero temperatures, there is a straight line relationship between air and water temperatures in terms of daily, weekly or monthly means (Fig. 3.5), even when the air and water temperature stations are some tens of kilometres apart (Crisp & Howson, 1982). The gradient of the calculated regression line is usually close to 1.0 (1.14 in the example shown in Fig. 3.5) and there is a positive or negative intercept (-0.3 in Fig. 3.5) that largely reflects differences in temperature between the air and water temperature recording stations. The calculated regression lines usually account for 90% or more of the variance of water temperature. There are, however, two general exceptions to these statements. When stream temperatures are measured close to the points of entry of major groundwater inputs (e.g. chalk or limestone springs) the relationship may be much less convincing and the gradient of the calculated regression may be significantly less than 1.0. Similarly, a lake or reservoir forms a

Table 3.6 Monthly mean water temperatures for Dockens Water (data for 1966–7 only), a small soft-water stream in the New Forest; for Bere Stream (mean of data for 1967 and 1968), a chalk stream tributary of the River Piddle; and for the River Frome (mean of data for 1962–8 inclusive). From Crisp *et al*., (1982).

	Dockens Water	Bere Stream	River Frome
Altitude (m.O.D.)	14	30	15
Mean discharge $(m^3 s^{-1})$	0.15	0.85	6.7
Jan.	3.2	9.2	6.5
Feb.	5.0	8.4	6.8
Mar.	6.9	9.8	7.4
Apr.	8.7	11.1	10.6
May	12.8	12.6	13.3
June	16.1	15.0	16.7
July	16.6	15.7	17.4
Aug.	15.4	14.7	17.0
Sept.	14.6	13.3	14.6
Oct.	11.1	11.9	11.9
Nov.	5.6	9.5	8.3
Dec.	5.4	8.4	6.7
Mean	10.1	11.6	11.4
Amplitude	13.4	7.3	10.9

Table 3.7 Monthly and annual means of daily water temperature ranges. Afon Cwm is a small moorland stream in mid-Wales (Crisp & Beaumont, 1995). The sites and periods of observation for the other three places are as in Tables 3.5 and 3.6.

	Afon Cwm	River Tees at Low Moor	Bere Stream	River Frome
Altitude (m.O.D.)	282	8	30	15
Mean discharge $(m^3 s^{-1})$	0.14	18.1	0.85	6.7
Jan.	1.79	0.18	0.72	1.10
Feb.	1.85	0.97	2.07	1.06
Mar.	2.59	1.30	3.64	1.46
Apr.	3.35	1.34	4.82	1.74
May	4.21	2.08	5.40	2.15
June	5.15	2.11	5.94	2.53
July	4.02	1.87	6.91	2.53
Aug.	3.92	1.62	5.29	2.27
Sept.	2.94	1.16	1.87	2.07
Oct.	2.26	0.84	2.51	1.31
Nov.	1.55	0.82	1.30	1.10
Dec.	1.49	0.81	1.06	1.24
Mean	2.93	1.26	3.46	1.71

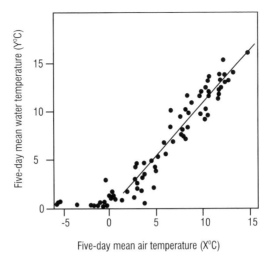

Fig. 3.5 Plot of five-days means of water temperature ($Y°C$) against five-day means of air temperature ($X°C$) for Mattergill Sike, a small upland stream in northern England. After Crisp & Howson (1982). The calculated regression line relating X and Y, omitting Y values below $0°C$, is $Y = 1.14X - 0.30$.

substantial 'heat sink' and both its water temperature and the temperature of water flowing out of it may lag substantially behind that of the air. In addition, some large bodies of standing water undergo 'thermal stratification' in summer with a layer of warm oxygenated water (the 'epilimnion') floating on top of a layer of cooler water often with reduced dissolved oxygen concentration (the 'hypolimnion') (*see* Fig. 5.9 on p. 111 for example). In general, the water temperature of a river immediately downstream of a lake or reservoir differs from that of a stream without such a heat sink and the relationship between water temperature and air temperature has a gradient significantly below 1.0 (0.59 immediately downstream of Kielder dam, northern England) and will account for a relatively small percentage of the variance of water temperature (<60% immediately downstream of Kielder dam). The water temperature in a river immediately downstream of a stratified reservoir depends upon the levels at which the released water is drawn-off, relative to the 'thermocline' (depth at which the epilimnion and hypolimnion meet). As it passes further downstream the temperature of the water approaches that of the air and the relationship with air temperature becomes closer. A similar situation occurs when, for example, heated effluent is released into a watercourse. For most practical applications, water temperature is recorded at a place well clear of the bank in a place of appreciable and, if possible, turbulent flow, by means of a probe that is shielded from direct solar radiation. This probably gives a reasonable picture of general temperature conditions within the main body of the stream. It is important to note, however, that temperature does vary spatially within streams. For example, even within a riffle in a large stony river in summer, the temperature may be a fraction of a degree (or more) higher close to the river margins than in mid-stream. Studies in a stream and river in southwestern England showed differences up to 1°C (usually less) between different parts of the cross-section (Webb *et al.*, 1996). In deeper rivers the temperature near the margins may be some degrees warmer in summer and cooler in winter than in the main course of the river; in the River Frome (southern England) temperature differences of up to 4°C were observed between different microhabitats (Garner *et al.*, 1998). There may also be small temperature differences between water within and water outside weed beds. These matters have been comparatively little studied.

An important aspect of spatial variation in temperature refers to the temperature of the water within the gravel of the river bed ('intragravel water'). This has relevance to the rate of development of salmonid eggs and alevins. There have been few published studies on this topic. In general, water temperatures within the gravel lag behind those in the water column and the lag increases with increasing depth in the gravel. This is true in terms of both daily and annual temperature cycles. In addition, the amplitudes of both daily and annual cycles reduce with depth within the gravel. Delays in the daily cycle and reduction of its amplitude are clearly seen in Fig. 3.6 for days of rising, relatively steady and falling temperatures in a typical trout spawning gravel in an English stream. The precise patterns of temperature within the gravel may, however, vary appreciably between stations as little as 0.5 m apart. There are also substantial differences between geographical regions. For example, Shepherd *et al.* (1986) studied gravels in British Columbia and Alaska and noted that the buffering effect of the gravel upon intragravel water temperature appeared to be complete at a depth of 10 cm. In contrast, the buffering was found to be not complete at 40 cm depth in the bed of a stream in northern England (Crisp, 1990b) or at a depth of 50 cm at a site in Oregon (Ringler & Hall, 1975). Such differences suggest that differences in stream bed structure (especially fines content and porosity) and in the rate and pattern of intragravel flow may be important influences. More recent studies have emphasised the influence of substratum characteristics on intragravel temperature patterns (Evans *et al.*, 1995) and the importance of upwelling and downwelling flows in gravel at different points within riffles (Evans & Petts, 1997).

Changing temperature modifies the physical properties of water. The viscosity (internal friction) of the water reduces as temperature rises and at 30°C has approximately half of the value that it has at 0°C. This has an influence on the rate of settlement of fine solids from suspension and also upon the energy required by fish to swim, though fishes reduce the effects of this 'viscous drag' by secreting mucus that reduces the friction between the body of the fish and the surrounding water. Although the density of water is close to $1.0 \, g \, ml^{-1}$ over a wide temperature range (Table 3.8) it has that precise value only at 4.0°C. At higher and lower temperatures the density of water is less than $1.0 \, g \, ml^{-1}$. This explains why in cold

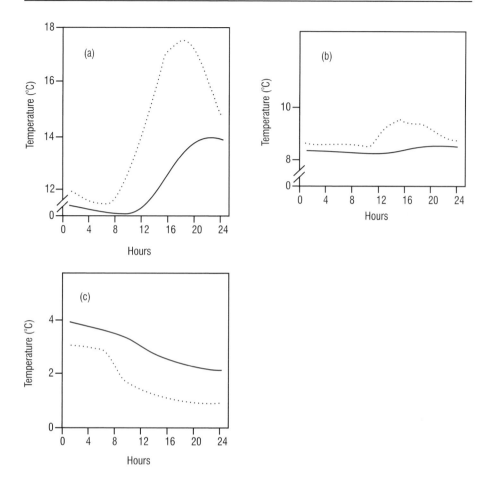

Fig. 3.6 Water temperatures throughout a day from midnight (0 hours) onwards at the gravel surface (dotted line) and at 20 cm depth within the gravel (solid line) for (a) a day of rising temperature (14 June 1986), (b) a day of relatively steady temperature (18 May 1986) and (c) a day of falling temperature (24 March 1986). Data from Carl Beck, Co. Durham, UK, after Crisp (1990b).

lakes during winter, the water at the bottom often has a temperature of 4.0°C while temperatures nearer to the surface are lower and ice may form at 0°C at the surface. The formation of surface ice on lakes and rivers acts as insulation and reduces the rate of cooling of the rest of the water column. In addition, substantial energy must be removed from water at 0°C to turn it into ice at the same temperature (the latent heat of freezing). For these reasons, in temperate regions it is very rare for even small streams to freeze completely. Even under relatively severe conditions there is usually free water in the stream bed and in small pockets beneath the ice. Unpublished observations in such streams suggest that small salmonids can survive these conditions for some weeks.

Table 3.8 Relationship of the density of water to temperature.

Temperature (°C)	Density of water (g ml^{-1})
0	0.99987
2	0.99997
4	1.00000
6	0.99997
10	0.99973
12	0.99953
16	0.99897
20	0.99823

There are two main ways in which water temperature can directly influence the biology of salmonids. The body temperature of salmonid fishes approximates that of the surrounding water and, therefore, temperature influences the rates of metabolism, growth, development and swimming speed. The relationships between these biological rates and temperature are usually curves and, on certain parts of those curves, small changes in temperature can cause disproportionate changes in the rates of the relevant processes. In some instances, for example the growth rate of trout and salmon on maximum rations (Elliott, 1975a; Elliott *et al.*, 1995; Elliott & Hurley, 1997), the rate increases with temperature to an optimum value and then decreases as temperature rises above that optimum. In addition to its effects on rates, changing temperature can switch on and off various behaviour patterns and biological activities. Use of the word 'switch' can, however, be a little misleading because the switching mechanisms are often clearly more complex than those implied by a standard domestic light switch. There are two main reasons. First, the exact temperature at which a given 'switch' will operate may be modified by the temperature regime to which the fish has previously been acclimated. Second, many of the 'switches' operate more like a dimmer switch than a simple on/off switch. This can be made clearer by reference to a type of device known as a 'thermal tolerance polygon'. These were first applied to fish by Brett (1952, 1956) who developed them for five species of *Oncorhynchus*. Subsequently, similar polygons have been produced for both of the species of *Salmo* (Elliott, 1981, 1991). A generalized example is shown in Fig. 3.7. The horizontal axis shows the acclimation temperature and the vertical axis shows the critical temperature (we might, perhaps, call this the 'switching temperature'). The body of the graph is divided into zones that can be tolerated by the species concerned in various respects. The inner edge of the shaded zone is the 'incipient lethal level' at which fish show signs of distress from high temperature but can survive for a considerable time. The outer edge of the shaded zone is the 'ultimate lethal level' and at this temperature death is almost instantaneous. Between these two levels there is a decrease in the survival time of the fish as temperature rises. In the centre of the figure are two smaller zones that show the temperature limits for

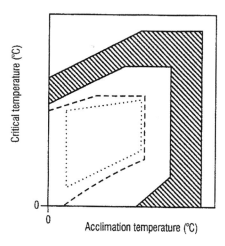

Fig. 3.7 Generalised thermal tolerance polygon for a salmonid fish species. The broken line marks the temperature limits for feeding, and the dotted line the limits for growth. The shaded portion is the critical range for survival. Its inner boundary marks the point at which fish may show distress but can survive for a considerable time ('incipient lethal level') and its outer boundary indicates the 'ultimate lethal level' at which death is almost instantaneous.

feeding and growth. The influence of acclimation temperature in modifying the exact value of each critical temperature is also apparent. In general, *Salmo salar* has a wider tolerance range, but at rather higher (by 1–3°C) temperature than does *S. trutta*. The possibility of genetic differences between populations in some temperature relationships is worthy of more study.

3.4   Gravel

The composition of stream bed gravel and various consequent effects are very important in the spawning and the intragravel life of salmonid fishes. They are components of a complex of factors that will be considered in more detail in Chapter 4. The present section is a general introduction to those aspects of the intragravel environment that are needed for an understanding of the biological implications.

In general, river bed gravels can be considered to have two main components. First, there is a framework of larger particles that support one another in an openwork structure. Second, the spaces in this framework are filled, to a greater or lesser extent, by finer material described as a 'matrix' (Carling & Reader, 1982). Consequently river gravels often have a bimodal size frequency distribution.

An introduction to the methodology of gravel sampling and analysis and to the treatment of the results is given by Shirazi *et al.* (1981); more detail, with reference to specific examples, is given by Platts *et al.* (1979). Gravel may be sampled by simple excavation, but freeze-coring methods give less disturbed samples and retain more of the finer material. The samples are usually analysed by passing them, either wet or dry, through a series of sieves of different mesh apertures. The fraction retained by each mesh aperture is then dried and weighed. Very often a standard sized set of mesh apertures is used which corresponds to the 'phi' scale. This is simply a scale of logarithms to the base 2, with sign changed (Table 3.9). Thus, a sieve with an aperture of 0 on the phi scale has a mesh aperture of 1 mm. Larger mesh sizes increase in steps by a factor of two on a negative scale so that apertures of 2 mm, 4 mm, and 8 mm are designated as −1 phi, −2 phi and −3 phi, respectively. Similarly, positive values of phi (1 phi, 2 phi, and so on) correspond to a series of mm sizes that decrease sucessively by a factor of two (0.5 mm, 0.25 mm, and so on).

Table 3.9 Sieve mesh aperture sizes in millimetres corresponding to various values on the phi scale.

phi value	mm
4	0.0625
3	0.125
2	0.25
1	0.5
0	1
−1	2
−2	4
−3	8
−4	16
−5	32
−6	64

One of the simplest ways to present the results of gravel analyses is to tabulate or plot the percentage of the sample by dry weight that falls within each size range. This has been done in Fig. 3.8 for samples from two chalk rivers in southern England. The bimodal distribution is shown clearly in both samples, with framework particles from about 2 mm to 64 mm and matrix at smaller sizes. Similarly, Carling & Reader (1982) showed that the matrix material in gravels from Teesdale (northern England) ranged in size from about 2 mm (−1phi) downwards. Another method of presentation is as a tabulation or graph showing cumulative percentage by weight with increasing mesh aperture. Figure 3.9 uses this method to compare single samples from a southern chalk stream and a Teesdale stream. Each sample has been split into an upper (0 to 100 or 160 mm depth) layer and a deeper layer and the compositions of the layers are shown separately. It is apparent that at both sites the proportion of finer materials is higher in the gravel from the deeper layers than from the upper layers. Plots of this type can be used to estimate the theoretical mesh apertures needed to pass

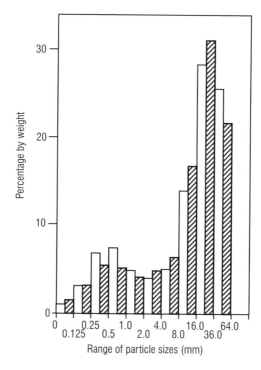

Fig. 3.8 Percentage of the total gravel sample within each range of particle sizes for gravel samples from two chalk rivers. Shaded bars are for the River Avon (Hampshire, UK) and open bars are for the River Piddle (Dorset, UK).

any given percentage of the total sample. For example, d_{50} is the theoretical mesh aperture that would pass exactly 50% of the sample and is described as the 'median grain size'. Similarly, d_{16} and d_{84} are the mesh sizes that would pass 16% and 84% of the sample, respectively. These two statistics can be used to compute the 'geometric mean grain size' (d_g), such that

$$d_g = \sqrt{d_{84}\ d_{16}} \quad \text{(Platts } et\ al.,\ 1979).$$ (5)

At this point the reader may well be puzzled by the plethora of statistics that seems to be used to describe gravel samples. The reason for this wide variety is that no single statistic describes adequately the composition of a given gravel and, therefore, a selection has to be used to even approach an adequate description. In addition to those already mentioned, the mean grain size and the 'sorting coefficient' (the standard deviation of the distribution) together with information on the 'skewness' and 'kurtosis' (as descriptions of the asymmetry of the distribution) are often quoted. A variety of transformations may also be used to achieve various statistical objectives.

The proportion of fine material, described by some authors as 'fines' and by

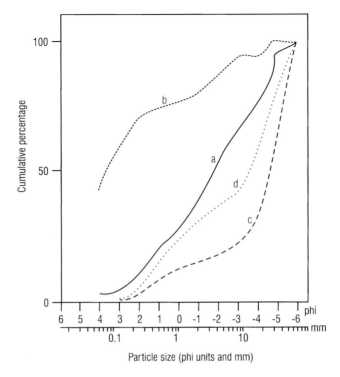

Fig. 3.9 Plots of cumulative percentage weight of gravel with increasing particle size. Replotted from Carling & Reader (1982) and from Beaumont, Ladle & Dear (unpublished). Data are for a stream in Teesdale, Co. Durham, UK for the uppermost 160 mm layer (solid line = a) and for deeper material (upper broken line = b); and for the River Piddle, Dorset, UK for the uppermost 100 mm (lower broken line = c) and for the 100–200 mm layer (dotted line = d).

others as 'sand', is considered to be important because of its influence upon water flow within the gravel and, hence, upon salmonid incubation. The percentage of fines can be expressed as P_x, which represents the percentage of the sample by weight that has a diameter of less than x mm. Unfortunately, different authors have used a wide variety of definitions of fines. Perhaps the most usual are $P_{1.0}$ and $P_{2.0}$ but commonly used sizes are quoted as ranging from $P_{0.083}$ to $P_{6.3}$ (Platts *et al.*, 1979) or from $P_{1.0}$ to $P_{5.0}$ (Milner *et al.*, 1981). This leads to a distinct lack of comparability between different published data sets.

In some circumstances a layer of larger stones may accumulate at the gravel surface to form an 'armoured' or 'pavement' layer. Such layers, in varying degree, reduce the likelihood of the finer material beneath being disrupted by spates (Carling & Reader, 1982).

An aspect of stream gravels that is of paramount importance to salmonid incubation is the flow of water through the gravel. This 'intragravel flow' occurs through the voids within the gravel, the space being quantified as the fraction (E) of a unit of cross-section of gravel that consists of void. Figure 3.10 shows three theoretical cross-sections with 100%, 45% and 15% void. The corresponding values of E are 1.0, 0.45 and 0.15, respectively. The void can be expressed as 'porosity' (λ), such that

$$\lambda = E/(1 + E) \quad \text{(Carling \& Reader, 1982)} \tag{6}$$

The 'apparent velocity' (V_A) of the intragravel flow (sometimes described as the 'superficial' or 'macroscopic' velocity) is the volume of water flowing per unit time through unit cross section of voids and solids. Cooper (1965) showed that V_A is a function of porosity, hydraulic head and permeability. Hydraulic head is, for example, the difference in water depth between the upstream and downstream ends of a gravel bed or portion of a gravel bed per unit of stream length along the line of flow (Pollard, 1955). This explains why salmonids usually construct redds at places where a hydraulic head is readily apparent – for example in gravel deposits at the foot of a pool where the pool merges into a riffle and there is an obvious slope to the water surface from the pool to the riffle. Permeability was defined by Coble (1961) as 'the capacity of gravel to transmit water'. This would be a factor additional to void space or porosity (as already defined) and related to the shapes and surface textures of the gravel particles and also to the viscosity of the water (hence to temperature). The 'seepage velocity' (V_S) (also described as 'true velocity' and 'pore velocity') is the velocity of flow per unit cross-section of pore space only. Although the exact value of V_A differs from pore to pore (Pollard, 1955), its mean value can be estimated from apparent velocity and the void fraction (E), as in the legend of Fig. 3.10. The seepage velocity through the egg pocket is one of the most important variables influencing the survival of salmonid eggs.

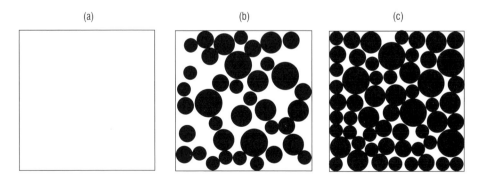

Fig. 3.10 Cross-sections of three theoretical bed materials. (a) is entirely hypothetical and refers to water with no gravel (100% void, E = 1.0). (b) is gravel with 45% void (E = 0.45). (c) is gravel with 15% void (E = 0.15). If apparent velocity in all three is V_A, then seepage velocity (V_S) is V_A in (a), $100\,V_A/45 = 2.2\,V_A$ in (b) and $100\,V_A/15 = 6.7\,V_A$ in (c). Similarly, the porosities (λ) will be 0.5, 0.31 and 0.13 for (a), (b) and (c) respectively ($\lambda = E/1 + E$).

It is important to recognize that the stream bed is a dynamic system. The coarse material may move during large spates and some of the matrix may move during spates and be deposited elsewhere during spate recession. Some fines may move during quite low flows. Relatively clean gravel frameworks can be rapidly in-filled by fines during low flows in those streams that carry substantial quantities of suspended solids (Carling & McCahon, 1987). A stream's supply of both coarse and fine material may be replenished by erosion of the stream banks and from other sources within the catchment and the rate of this supply may vary with land use and with other human activities.

Chapter 4
Environmental Requirements and Limitations

(ideal homes)

Summary

The precise environmental requirements may vary between species but show similar general patterns. Water temperature affects all life stages via such variables as survival, growth, development, metabolic rate and swimming speed. Similarly, water quality is important to all life stages especially through toxins, suspended solids, pH values and dissolved oxygen concentrations. The precise lethal or switch values of these variables may differ between life stages within a species. Water velocity, depth, wetted area and habitat structure are important to juveniles and to resident and migratory adults through their influence upon habitat availability and social interactions. Spawning movements can be related to river flow and habitat structure, with special reference to hiding places and obstructions to movement. Successful spawning and incubation require suitable gravel structure, intragravel flow and intragravel oxygen concentration. These are, in turn, influenced by river bed conformation, river flow, gravel composition and various hydrological and hydraulic effects. Egg and alevin survival can be harmed by infilling of gravel interstices by silt and by gravel movement during spates. Some of these interrelationships can be partially represented by simple mathematical models but others are much less well understood at functional and quantitative levels.

The details of the environmental requirements of trout and salmon vary between species (possibly also between populations) and between life cycle stages. To consider the requirements, it is therefore necessary to divide the life cycle into several life stages. We will consider three stages, namely: intragravel, juveniles and smolts plus adults and spawning. This still leaves a rather 'messy' situation because, for example, the habitat of the intragravel stages is determined largely by the behaviour of the spawners and this leads to some overlap when we consider the requirements of these two stages.

4.1   Intragravel stages (redds under the bed)

First we will consider those factors likely to influence intragravel survival, to indicate the mode of action of each one and, as far as possible, to quantify their effects. As will become apparent, however, many of these factors interact and this makes it difficult to assess their effects quantitatively.

4.1.1   Temperature

This is the main factor determining the rate of development and duration of the intragravel stages. It is important to predict, with reasonable accuracy, the progress of intragravel development from oviposition onward for two reasons. First, such predictions are essential for rational conduct of field and laboratory experiments on intragravel stages. Second, these predictions give us an estimate of the progress of embryonic development and of the timing and duration of the periods of maximum vulnerability to various types of risk. For example, the greatest risk from mechanical shock occurs between the time of oviposition and the time of median eyeing and suitable predictions of eyeing date will indicate the approximate point in time at which this particular risk diminishes.

The literature contains a number of mathematical models (equations) for predicting from temperature ($T°C$) the number of days (D_2) required to reach median hatch (i.e. for 50% of a given batch of eggs to hatch) for eggs of a variety of salmonid species (Table 4.1; Fig. 4.1). More than one model is cited for some species but it is worth noting that all of the models given for any one species give similar predictions over the temperature range for which each one is relevant. The model for *S. salar* is based on relatively few data points and a rather narrow range of temperatures but, despite this, has been found to work well even at temperatures below 1°C (Wallace & Heggberget, 1988). The quantity $100/D_2$ is the percentage daily increment towards median hatch. From the date of oviposition onward, daily mean temperatures can be used to calculate the percentage increment for each day. The predicted day of median hatch (when exactly 50% of the surviving eggs will have hatched) will be the day upon which the cumulative sum of daily increments reaches 100%. If D_1 is taken as the number of days to median 'eyeing' of the eggs (a somewhat subjective judgement) then

$$D_1 = b \, D_2 + a \tag{7}$$

when $b = 0.455 \pm 0.102$ (95% CL) and $a = 5.0 \pm 12.7$ (Crisp, 1988b). Similarly, if D_3 is taken as the number of days to median swim-up, then

$$D_3 = b_1 \, D_2 + a_1 \tag{8}$$

Table 4.1 Models for predicting the time of median hatch (D_2, days) from temperature (T°C) for seven species of salmonid fish. The sources of the information are: 1 = Crisp (1981); 2 = Humpesch (1985); 3 = Jungwirth & Winkler (1984). The equations of Crisp (1981) are based on data gathered from the literature, whereas those of Humpesch and Jungwirth & Winkler are based on original data. For further discussion of other models *see* Crisp (1992).

Species	Equation	Temperature range (°C)	Source
Salmo trutta	$\log_{10}D_2 = [-13.93061\log_{10}(T + 80.0)] + 28.8392$	1.9–11.2	1
S. trutta	$D_2 = 281\ T^{-0.84}$	1.4–15.2	2
S. trutta	$D_2 = 746.0/(T + 0.5323)^{1.2233}$	about 5.0–13.0	3
S. salar	$\log_{10}D_2 = [-2.6562\log_{10}(T + 11.0)] + 5.1908$	2.4–12.0	1
Salvelinus alpinus	$D_2 = 206\ T^{-0.63}$	1.0–8.0	2
S. alpinus	$D_2 = 476.76/(T-0.1314)^{1.0435}$	about 4.0–11.0	3
Hucho hucho	$D_2 = 2647.4/(T + 3.222)^{1.7865}$	about 4.0–15.0	3
Oncorhynchus mykiss	$\log_{10}D_2 = [-2.0968\log_{10}(T + 6.0)] + 4.0313$	3.2–15.5	1
O. mykiss	$D_2 = 456\ T^{-1.14}$	10.0–19.0	2
O. tshawytscha	$\log_{10}D_2 = [-1.8126\log_{10}(T + 6.0)] + 3.9166$	3.6–18.1	1
Thymallus thymallus	$D_2 = 459\ T^{-1.37}$	3.5–16.2	2
T. thymallus	$D_2 = 6484.6/(T + 5.103)^{2.099}$	about 3.0–15.0	3

when $b_1 = 1.660 \pm 0.103$ and $a_1 = 5.4 \pm 10.05$ (e.g. Fig. 4.2). The relationship between D_1 and D_2 was based on data for *S. salar* only, whereas the relationship between D_3 and D_2 was based on combined data for both species of *Salmo* and four species of *Oncorhynchus*. Both models were tested on *S. trutta*, *S. salar* and *Salvelinus alpinus* and were found to work well. In many practical applications it is adequate to assume that

$$D_1 = 0.5\ D_2 \tag{9}$$

and that

$$D_3 = 1.7\ D_2 \tag{10}$$

or that

$$D_3 = 1.5\ D_2 \tag{11}$$

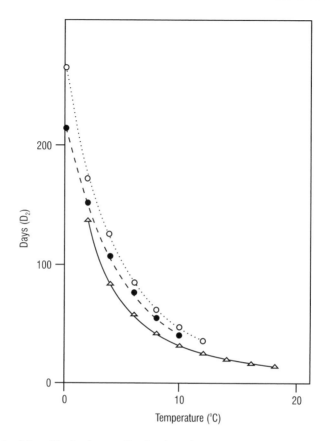

Fig. 4.1   Plots of time (D_2 days) to median hatch against temperature for eggs of *S. salar* (open circles), *S. trutta* (filled circles) and *O. mykiss* (triangles).

under hatchery conditions and

$$D_3 = 2.0 \, D_2 \tag{12}$$

in natural redds (Crisp, 1992). In summary, we can predict the date of median hatch from water temperatures by the use of relationships between D_2 and T that have been developed for each individual species. Approximate dates of median eyeing (D_1) and median swim-up (D_3) can be predicted from the D_2 and T relationships for individual species by use of more general relationships that appear to be common to most species of *Salmo* and *Oncorhynchus* and to at least one species of *Salvelinus*.

These relationships have been developed and tested using a variety of species from a number of different provenances and this suggests that there is probably little or no between-populations variation within each species. Brännäs (1988) suggested, however, that there may be differences in development rates between Baltic and Canadian populations of *S. salar*. Donaghy & Verspoor (1997) found

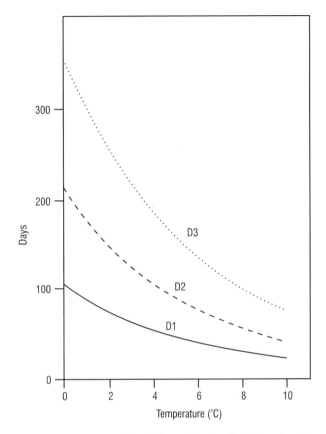

Fig. 4.2 Plots of days to median eyeing (D_1), median hatch (D_2) and median swim-up (D_3) against temperature (°C) for *S. trutta*.

statistically significant differences in hatching time between *S. salar* eggs from two Scottish rivers but the differences in median hatching dates were small (one to three days). Berg & Moen (1999) observed similar differences between populations of salmon in Norway but, though statistically significant, the observed differences were small (six days or less). Practical experience suggests that the various predictions are usually accurate to within ± 7 days. Exceptions do occur and discrepancies of up to 30 days may result from delays caused by low temperatures just before hatching (Crisp, 1988b) or as a result of sub-lethal mechanical shock (Crisp, 1990a). When eggs are stressed in various ways premature hatching may occur. Additional factors that can exert some influence on embryonic development include incident light level (Bieniarz, 1973; Hamor & Garside, 1975) and dissolved oxygen concentration (Hamor & Garside, 1976). The date of swim-up can be influenced by gravel composition and may occur earlier (hence, at a less well developed stage) in gravel-free rearing units than in natural gravels.

A more complex model to predict swim-up time for sea trout is given by Elliott & Hurley (1998b). It is important to note that this refers to eggs incubated in ideal media at temperatures between 1.5 and 10.5°C. Predictions based on this model may be misleading if applied to material incubated in sub-optimal media and/or subjected to temperatures around 0°C at the time of predicted hatch.

We have already noted in Chapter 3 that the temperature of intragravel water may differ from that of the free water within the stream. Several authors have commented on the possible relevance of this to salmonid incubation but there have been few detailed studies. In a stream in northern England (*see also* Fig. 3.6) predicted dates of median eyeing and swim-up of *S. trutta* were only one to three days earlier at 20 cm depth in the gravel than at the gravel surface but the predicted date of median hatching was up to 11 days earlier at 20 cm than at 0 cm depth (Crisp, 1990b). In contrast, a study in streams in southern England showed that development in all embryonic stages was more rapid within the gravel than at the gravel surface. The predicted date of swim-up at 25–30 cm depth in the gravel was between 5 and 22 days earlier than at the gravel surface (Webb & Clarke, 1998). These findings show that there is considerable variation between sites as a result of differences in oviposition dates, timing of spawning relative to the annual temperature cycle and the properties of the gravel.

The lethal temperature limits and the temperature ranges within which > 50% survival of eggs to the point of hatch can be expected are summarised in Table 4.2 for six salmonid species. The two *Salmo* species and *Salvelinus alpinus* show good survival to hatch (about 100%) down to 0°C. At higher temperatures the survival rate decreases and becomes zero at the upper lethal limit. There is, therefore, a dimmer switch action towards the upper lethal limit (Fig. 4.3a). In contrast, the lower lethal limit for *Hucho*, *Thymallus* and *Oncorhynchus mykiss* is above freezing point and survival is represented by a domed curve. This is equivalent to having a dimmer switch at each end of the range (Fig. 4.3b). Humpesch (1985) produced similar graphs for six species and for three of them he used fish from two different provenances. For *S. trutta* (Fig. 4.3a) there was little difference

Table 4.2 Approximate upper and lower lethal temperature limits and the range in which there is > 50% survival to hatch, for eggs of six species of salmonid fish. Sources: 1 = Humpesch (1985); 2 = Gunnes (1979).

Species	Lethal limits (°C)		Temperature range to give > 50% survival to hatch	Source
	Upper	Lower		
Salmo trutta	15.5	< 0	0–11.0	1
S. salar	> 12.0	< 0	0–< 12.0	2
Salvelinus alpinus	12.5	< 0	0–7.5	1
Hucho hucho	15.5	1.5	5.0–13.0	1
Thymallus thymallus	18.5	3.0	4.1–7.5	1
Oncorhynchus mykiss	> 19.0	< 3.0	5.0 (?)–15.0 (?)	1

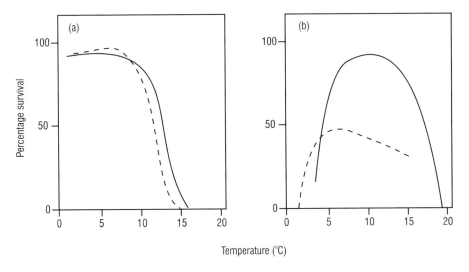

Temperature (°C)

Fig. 4.3 Relationship between percentage survival to hatch, and incubation temperature, for two provenances of *S. trutta* (a) and two provenances of *O. mykiss* (b), simplified from Humpesch (1985).

between the two batches. For *O. mykiss* (Fig. 4.3b) and *Thymallus*, however, there was evidence of differences between provenances. More data are therefore needed before we can be sure whether or not there are differences between populations and/or whether there is any acclimation effect with regard to the survival of salmonid eggs at different temperatures. Until such information is available, the values given in Table 4.2 can, at least, be taken as an approximate guide.

In early studies on the incubation of trout eggs, Gray (1928) noted that the relative amounts of yolk used for metabolism and for elaboration of tissue varied with incubation temperature. Lower incubation temperatures reduced the proportion of the yolk used for metabolism and gave larger fry. The size of fry may influence their survival (Elliott, 1984a). Gray also suggested that embryonic development and hatching had different lower temperature thresholds with the latter higher than the former. This might explain why hatching is delayed when very low temperatures occur at the time of predicted hatch.

4.1.2 *Chemical effects*

The eggs and alevins of trout and salmon are vulnerable to low pH in the intragravel water. Water at pH 3.5 is lethal to trout and salmon eggs within ten days but no effect of pH alone can be shown at pH 4.5 and above (Carrick, 1979). Nevertheless, pH values above 4.5 can have a lethal effect in association with high concentrations of monomeric aluminium or some heavy metals (Table 3.4). Tests at pH 4.5 (Brown & Lynam, 1981) showed that a calcium concentration of $10 \, mg \, l^{-1}$ was needed for survival of newly fertilized eggs of *S. trutta* but that eyed eggs could survive in deionized water that had been acidified with sulphuric acid, but to which no other ions had been added. Concentrations of $1 \, mg \, l^{-1}$ each of sodium and calcium ensured survival to hatching from the eyed egg stage and throughout the alevin phase. At pH values below 4.5 the action of the hatching enzyme (chorionase) of *S. salar* is blocked and there may be death at the time of hatch (Waiwood & Haya, 1983). A similar mechanism is likely to apply to the eggs of other salmonids. So far as intragravel stages are concerned, the general salmonid guideline (Table 3.4) that low pH is harmful at values below 5.0 and lethal at values below 4.0 appears to be realistic.

Geertz-Hansen & Mortensen (1983) showed that naturally occurring concentrations of iron in some Danish streams at pH 6.55 to 7.05 caused increased mortality of *S. trutta* eggs and alevins.

4.1.3   *Mechanical shock*

Sensitivity to mechanical shock (impact, pressure, vibration) has been studied in several species of *Oncorhynchus*. Smirnov (1955, 1975) and Ievleva (1967) shocked eggs of *O. keta* and *O. nerka* by means of a vibrator and noted a rapid increase in sensitivity within minutes of fertilization. This was followed by a period of reduced sensitivity and then a further increase in sensitivity that lasted up to the eyed stage. In contrast, Jensen & Alderdice (1983) subjected the eggs of *O. kisutch* to an impact of known energy at various stages of development. They observed stepwise changes in sensitivity. The eggs showed little sensitivity to shock during the first ten minutes after fertilization or after the first fifteen days at 10°C (probably close to the time of median eyeing). A similar general pattern was found in *O. mykiss* (Dwyer *et al.*, 1993). For eggs incubated at 10.4°C mechanical shock gave high mortality rates (70 to 100%) between days two and eight (6.5 to 26% D_2 – see Table 4.1) but the rate of mortality then reduced and became negligible in eggs shocked by day 14 (45.8% D_2 and a day or two before predicted median eyeing). The effects of the type of shock likely to be inflicted by electrofishing over redds was also tested. Its effects were less than those of mechanical shock and they peaked at a mortality rate of about 60% for eggs shocked on day eight (26% of D_2) and became negligible by day 12 (39% D_2). Under field conditions a brief period of resistance to mechanical shock might be expected immediately after fertilization in order that the eggs might not be killed during the burial process. There are some indications (e.g. Jensen & Alderdice, 1983) that the sensitivity of salmonid eggs to shock might vary between species. There is therefore a general consensus that sensitivity develops at or soon after fertilization and reduces markedly around the time of eyeing. Less thorough studies (Crisp, 1990a) indicate a similar pattern for *Salmo* eggs to that quantified in detail by Jensen & Alderdice (1983) for *O. kisutch*. There is also some evidence that death from mechanical shock may sometimes not occur until up to 60 days after the application of the shock and also that sub-lethal shock may cause delays in hatching by up to two weeks. Mechanical shock can be administered to salmonid eggs in the field in several ways. The passage of people or vehicles over redd sites can be important. Substantial mortality of eggs and alevins can be caused by anglers wading over redd sites (Roberts & White, 1992) and it therefore requires little imagination to visualise the damage that can be done by the passage of tractors or civil engineering machines. Eggs may also be released into the water when redds are disturbed by spates or by human activity. Trout eggs allowed to drift 10 m in an experimental stream channel suffered approximately 50% mortality when drifted at 10 to 20% development to median hatch but appeared unharmed when drifted at 60 to 70% development to median hatch (i.e. after eyeing) (Crisp, 1990a).

Eggs are sometimes used for stocking. For best results the transportation and planting of eggs in artificial redds should occur after eyeing when sensitivity to mechanical shock is minimal.

4.1.4 Egg burial depth

A large number of studies on burial depths of salmonid eggs was reviewed by De Vries (1997). Depths of 0–80 cm have been recorded, though values between 0 cm and 50 cm are more usual. The reviewer drew attention to the wide variety of methodologies and definitions used and noted that comparatively few of the observations on redds were accompanied by details of the lengths of the female fish responsible for their construction. Early work on *Salmo*, *Oncorhynchus* and *Salvelinus* species (Greeley, 1932; White, 1942; Hardy, 1963) suggested that burial depth was probably more dependent on the size of the female fish than on her species and later studies tend to endorse this view. It is likely that gravel composition and water velocity also play a part but, as will be seen, these may also be related to female size. A general guide to burial depths for the two species of *Salmo* and for seven species of *Oncorhynchus* is given in Fig. 4.4. De Vries referred to the studies of Ottaway *et al.* (1981) and Crisp & Carling (1989) (the latter an expansion of the former) for *Salmo* species, and to van den Berghe & Gross (1984) and Holtby & Healey (1986) for *Oncorhynchus*, for preliminary quantitative relationships between egg depth and fish size. Mean egg depth (D cm) for *Salmo* species in the UK can be predicted from the equation

$$D = bL + a \tag{13}$$

where L is female fish length (cm), a ± 95 % CL is 2.4 ± 7.53, and b ± 95% CL is 0.262 ± 0.098 for fish of 24–74 cm length. Most eggs are to be found within 3 cm of the mean depth below the general gravel surface (Crisp & Carling, 1989). This relationship accounts for over 60% of the variance of egg depth for sites in typical

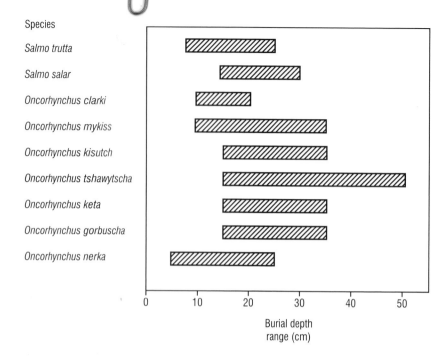

Fig. 4.4   Ranges of egg burial depths likely to be covered by the egg pockets of two species of *Salmo* and seven species of *Oncorhynchus*. Note that *S. trutta* and *O. mykiss* include both resident and sea-going forms and *O. nerka* includes 'kokanee' (the land-locked form). Derived from De Vries (1997).

stony streams and rivers in southwestern Wales and northeastern England. Occasionally the occurrence of, for example, a large flat stone in the pit, may result in shallower egg burial than would be expected from the size of the female fish. No correlation between fish size and burial depth could be detected for chalk streams in southern England where the eggs of fish from 45–84 cm length were buried at mean depths between 8.75 cm and 27.0 cm. This may be related to the occurrence of a 'cemented/concreted' layer at 5–10 cm depth in the beds of many English chalk streams. Within these layers, adjacent gravel particles can be bound together by deposits of calcium carbonate (Scott & Beaumont, 1994). Van den Berghe & Gross (1984) give the equation

$$D = 0.411\ L - 10.44 \tag{14}$$

for *O. kisutch* of 46–74 cm length ($r^2 = 0.61$). Over the range 50–75 cm this gives rather shallower predicted egg depths than does the relationship for *Salmo* species, particularly for the smaller fish. This difference may, however, be a reflection of methodology. In the *Salmo* study, egg depth was determined by the examination of frozen cores, whereas the depth of the pit of the completed redd was taken as an estimate of egg burial depth by van den Berghe & Gross. Should female coho salmon show the same tendency as female *Salmo* to cut randomly round the edges of the pit after the completion of spawning (*see* Chapter 2), then the pit depth is likely to underestimate the egg burial depth.

4.1.5 *Hydrological effects*

Patterns of discharge fluctuation influence stream depth and width, water velocity, the hydraulics of intragravel flow, the transport and deposition of fine silt and movements of the stream-bed gravel. Changes in wetted area and water velocity will give changes in the area available for salmonid spawning, and decreases in wetted area after spawning may lead to exposure of redds to desiccation or freezing. Changes in intragravel flow as a result of modified hydraulic head and/or silt deposition in the gravel voids may affect oxygen supply to intragravel stages.

   Large spates may modify stream courses and the riffle and pool sequences and the movement of gravel beds may cause washout of salmonid eggs and alevins. Harris (unpublished thesis, Liverpool University 1970) noted that 0 to 58% (mean 27%) of sea trout redds were washed out by spates in some tributaries of the River Dyfi (Wales). Washout can cause the death of intragravel stages through physical damage during the disruption of the redd, physical shock and predation during drifting and through subsequent deposition in sites unsuitable for continued development. Washout of redds is considered likely to be a major cause of mortality in some Pacific salmon populations (Wickett, 1952; Gangmark & Bakkala, 1960; Lister & Walker, 1966).

   Typical egg burial depths for *Salmo* are 5–30 cm (Fig. 4.4) and burial depth is likely to have some bearing on the likelihood of egg washout. At present, it is generally agreed that there is not an adequate basis of physical science for sound prediction of the locations, areas or depths of gravel that are likely to be disrupted by a spate of given size in any given river section. A correlation between egg-to-fry mortality and peak instantaneous discharge during the incubation period was found by Holtby & Healey (1986) for *O. kisutch* in Carnation Creek, Vancouver Island, and was attributed to egg washout during spates. Elliott (1976a) was able to relate the concentration of drifting *S. trutta* eggs to discharge in two English streams by means of semi-logarithmic equations. The intercepts in the equations differed from year to year and were assumed to reflect varying densities of eggs from one year to another. Both of these studies present some problems because bed movement in gravel-bedded streams can be expected to be a function of previous spate history (and possibly other factors) as well as of the discharge in individual spates (or the largest individual spate). In addition, the Elliott study implies negligible washout, relative to initial egg numbers, otherwise some depletion effects through the course of the incubation period might be expected to upset the relationships. A little information is available from empirical investigations. Burial and subsequent excavation of colour-coded artificial eggs at 5, 10 and 15 cm depth in a typical *Salmo* spawning site showed that most eggs (90%) at 5 cm, variable numbers at 10 cm and a few at 15 cm were washed out by the sort of spate that occurs several times per year (Crisp, 1989). A spate of between 10 and 20 years' return period, however, washed out almost all eggs at 5 and 10 cm and over 40% at 15 cm. This is consistent with a very approximate guideline that the disturbance of bed material in a gravel bedded stream typically occurs to a depth of one mean grain size (Carling, 1983), though disturbance to greater depths in exceptionally large spates cannot be ruled out. This study also showed that some of the eggs not washed out were displaced downstream within the gravel by about 85 cm and, had they been real eggs, this could have caused some damage by crushing and/or mechanical shock.

4.1.6 Bed conformation and hydraulics

The importance of hydraulic head in relation to intragravel flow has already been noted (*see* 3.4 Gravel) and a number of authors have described a typical redd site as being at a point where there is a marked change in the hydraulic head above the gravel. Such sites are found at the downstream ends of pools where water accelerates over the gravel to enter a riffle (Stuart, 1953a) and also within some riffles, though some small trout may spawn in quite small gravel patches between large stones. Stuart (1953b) and Cooper (1965) used dyes to trace intragravel flow and demonstrated increased flow through gravel bars at the feet of pools and also where large stones protruded above the gravel surface. The general consensus (e.g. Vaux, 1962) is that intragravel flow depends mainly upon hydraulic head, gravel permeability and bed surface profile. It can be argued (e.g. Peterson, 1978) that the typical form of a newly constructed redd will tend to increase intragravel flow but, in many upland streams prone to severe spates, the typical contours of a redd may disppear rapidly (e.g. Ottaway *et al.*, 1981).

4.1.7 Oxygen supply, 'waste disposal' and gravel composition

Minimum dissolved oxygen concentrations (mg 1^{-1}) that can give 'incipient sub-lethal responses' in embryos of *S. salar*, based on work by Lindroth (1942), Hayes *et al.* (1951) and Wickett (1954), are given by Davis (1975). For early eggs these oxygen concentrations varied from 0.53 mg 1^{-1} at 5.5°C to 2.17 mg 1^{-1} at 10°C. Eggs close to hatching required a much higher concentration, which varied from 4.06 mg 1^{-1} at 5.0°C to 7.00 mg 1^{-1} at 17.0°C. After hatching the required oxygen concentration decreases (Hayes *et al.*, 1951). The minimal concentrations given by Davis apply when the rate of flow of water past the eggs is sufficient to ensure that the effects of oxygen consumption by each embryo upon the concentration around it or other embryos are negligible. The rate of flow of water past eggs/alevins buried in natural gravels is, however, often rather low and it is then necessary to consider the combined effects of oxygen concentration and intra-gravel flow (i.e. oxygen supply rate) and compare these to the rate at which oxygen is being consumed by the eggs/alevins. Some of the complexities of the interrelationships of factors can be seen in Figure 4.5. In this connection it is important to note several points. First, the oxygen requirements of salmonid eggs depend upon egg weight, stage of development, water temperature, and possibly species. Second, the rate at which an egg can take up oxygen depends upon its ratio of surface area to volume (Wickett, 1954) and egg size varies considerably within salmonid species with fish size (Elliott, 1984a; Crisp *et al.*, 1990; L'Abée-Lund & Hindar, 1990) and trophic status (Bagenal, 1969). Third, the survival of a given egg depends upon its spatial relationship with other eggs and, hence, upon the distribution of eggs within the egg pocket (Wickett, 1954). These complexities (coupled with differences in definition, failures of some authors to give egg

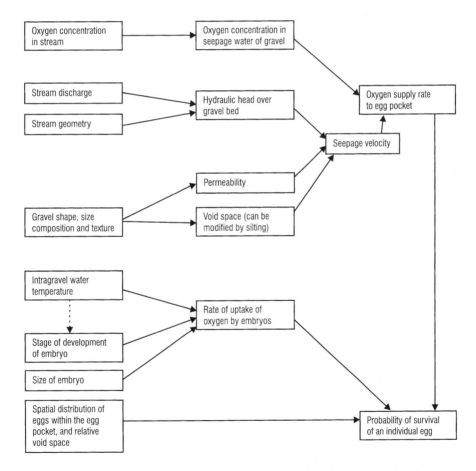

Fig. 4.5 Schematic representation of the main factors affecting the intragravel environment and those related effects that link them, via oxygen supply rate, to the survival of an individual salmonid egg. After Crisp (1989).

weights, stages of development or water temperature, technical difficulties in sampling interstitial water and the use of apparent velocity rather than seepage velocity) probably explain why no clear and detailed statement on the oxygen supply rate required by salmonid eggs is possible at present. Crisp (1996a) concluded that the balance of evidence suggested that high egg survival can be expected, even at the time of high demand around hatching time, if oxygen concentration remains above $7.0\,mg\,l^{-1}$ with temperatures below $12.5°C$ and apparent velocity at or above $100\,cm\,h^{-1}$. This latter value probably corresponds to a seepage velocity of about $250–650\,cm\,h^{-1}$. Lower values of concentration and seepage velocity may be satisfactory at lower temperatures and earlier stages of development.

Oxygen deficit may also have sub-lethal effects, including reduced growth, reduced yolk conversion efficiency, premature hatching, reduced size at hatching and various morphological changes (Garside, 1959, 1966; Hamor & Garside, 1977). Some of these effects may occur some time after the occurrence of hypoxia (Mason, 1969).

The main metabolic waste product of salmonid eggs and alevins is ammonia, which is toxic to salmonids (Table 3.4). The intragravel stages are dependent upon intragravel flow to carry away this waste. This aspect has been little studied but it is likely that when seepage velocity is low and eggs/alevins are densely packed, the contribution of each individual increases the probability of a toxic concentration being received by the next individual 'downstream'.

In terms of oxygen supply rate, the eggs of salmonids are heavily dependent upon the choice of redd site made by the female parent. If conditions become unsuitable, they cannot move to a more congenial place. In contrast, alevins do have ability to move through the gravel (Heard, 1991), but it is not clear to what extent this ability is used to move in such a way as to optimise the environmental conditions.

Gravel composition is important because of its influence upon oxygen supply rate (Fig. 4.5) and also because it can influence the success of emergence from the gravel and the time of swim-up. The literature contains a large number of papers detailing experiments and observations on the relationships between gravel composition and survival to swim-up. The general patterns that emerge from these studies and which are common to most, can be well illustrated by reference to two published data sets. Olsson & Persson (1986) examined the relationship between gravel size and survival to swim-up of *Salmo trutta* eggs and alevins. Eggs were planted within 24 hours of fertilization in artificial redds filled with homogeneous gravel of five different sizes from 1.5–32.0 mm. The numbers of alevins surviving to swim-up increased with gravel size up to 18.0 mm and there was a levelling off or a slight fall in survival rate between 18.0 mm and 32.0 mm gravels (Fig. 4.6). In essence, this experiment was a test of the effect of larger (framework) gravel in the absence of finer (matrix) material. The results indicate an optimum particle size in the region of 20–30 mm and this is in good agreement

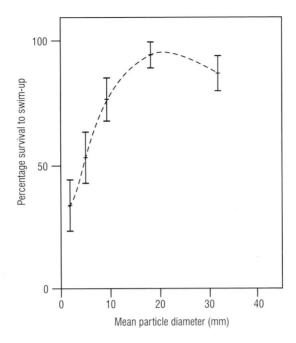

Fig. 4.6 Percentage survival to swim-up of *S. trutta* incubated in homogeneous gravels of five different mean particle diameters. 95% CL are shown. Plotted from data in Olsson & Persson (1986); line fitted by eye.

with field observations on the gravel sizes selected by spawning salmonids (Shirazi *et al.*, 1981; Fluskey, 1989; Crisp & Carling, 1989). Later experiments (Olsson & Persson, 1988) examined the effects of increasing quantities of fines (sand of 0.75 mm diameter) in a standard framework composed (before addition of fines) of 75% (volume) of 18 mm gravel and 25% (volume) of 4.8 mm gravel. The results (Fig. 4.7) show a very steep decrease in survival at sand contents between 10% and 20%, by volume. Peterson & Metcalfe (1981) performed similar experiments on *S. salar* and obtained very similar results. There is also evidence that at higher concentrations of fines there may be premature emergence. For example, Olsson & Persson (1988) found that at a sand content of 10% or less, swim-up occurred after 98 days, whereas at 20% and 40% sand content swim-up occurred on average at 70 days and 55 days respectively. The individuals that emerged prematurely were smaller and had more residual yolk than the others.

Despite the wealth of published material, it is difficult to derive firm quantitative recommendations regarding gravel composition and fines content. There is a general lack of comparability between different experimental studies and this arises from three main sources. First, there is a wide variety of different definitions of 'fines'. Second, few of the published accounts contain adequate information on the stage of development of the eggs at the start of the experiment, the

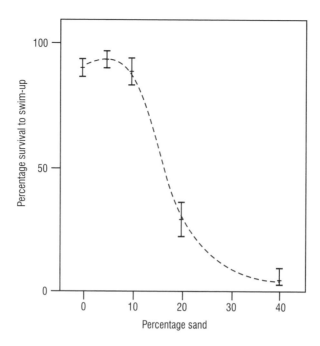

Fig. 4.7 Percentage survival to swim-up of *S. trutta* incubated in a standard gravel mix to which different percentages, by volume, of 0.75 mm sand had been added. 95% CL are shown. Plotted from data in Olsson & Persson (1988); line fitted by eye.

depth at which they were buried and water temperature during the experiment. All of these variables are very relevant. Third, the experimental design generally fails to distinguish between the two different possible effects of fines content, namely its effects upon oxygen supply via intragravel flow (the 'incubation' effect) and its effect in hampering emergence from the gravel (the 'emergence' effect). This is true of three studies on *S. trutta* (Witzel & MacCrimmon, 1983; Olsson & Persson, 1986, 1988) and four others on *S. salar* (Peterson & Metcalfe, 1981; MacCrimmon & Gots, 1986; Marty *et al.*, 1986; O'Connor & Andrew, 1998). Field studies have their own difficulties with regard to definitions but they are also beset by severe technical problems. In particular it can be argued (e.g. Chapman, 1988) that no meaningful conclusions can be drawn without good measurements of conditions within the egg pockets and this presents severe difficulties in both methodology and cost.

There have been relatively few studies designed specifically to examine the effects of fines on the process of emergence. Most of these have been concerned with *Oncorhynchus* species. Field studies (Koski, unpublished thesis, Oregon State University, 1966) demonstrated a much lower emergence success through fine gravels than through coarse gravels in *O. kisutch*. In experimental studies on *O. mykiss* and *O. kisutch*, Phillips *et al.* (1986) found emergence rates of 90–100% when sand (1.0–3.0 mm) content was less than 10%; 60 to 80% with 10–20% sand;

and only 20–30% when sand content was above 60%. In addition, several authors have noted the somewhat smaller size of alevins/fry emerging from gravels with high percentages of sand than from those with lower percentages. These observations refer to the most typical situation in which the fines are spread within the framework particles. In some natural situations, however, a sand layer may settle and 'float' on top of the framework or may settle within the uppermost layer of the framework. Bams (1969) observed alevins of *O. nerka* pass through such sand layers by vigorous 'butting' behaviour. Experiments on *S. trutta* and *S. salar* (Crisp, 1993c) gave somewhat inconclusive results but did not show any clear-cut effect of sand layer thickness (0 cm, 2 cm, 4 cm and 8 cm) upon emergence success or timing for either species. It was clear that alevins of both species could successfully penetrate sand layers of up to 8 cm thickness.

In a detailed and closely reasoned review, Chapman (1988) commented that 'one cannot, with the existing information on survival of embryos and alevins in the redds of large salmonids, predict survival quantitatively and with known accuracy on the basis of physical factors measured in field and laboratory studies'. A decade later, this still appears to be a fair assessment of the position. Despite this, as an interim measure it is useful to derive some working guidelines from the available literature. In very broad terms we can suggest that a satisfactory incubation gravel is likely to have a mean grain size of 20–30 mm and contain less than 10–20% of fines (particles of less than 1.0 mm). These guidelines should be used cautiously and in conjunction with the criteria that intragravel dissolved oxygen concentration should ideally be 7.0 mg l^{-1} or more, and seepage velocity should probably be at least 250–650 cm h^{-1}. Observed porosity values in typical UK spawning gravels are 0.16–0.35 (Crisp & Carling, 1989).

Crisp & Carling (1989) compared percentages of fines (< 1.0 mm) in gravel samples taken from the surface layers of *Salmo* redds and from undisturbed gravels around the redds and beneath the redd tails. On average, a smaller content of fines occurred in the disturbed gravel of the redds than elsewhere and this implies some cleaning of the gravel during redd construction. However, the range of values and the 95% CL of the means were large, and it was concluded that in general there was no convincing evidence of changes in grain size as a consequence of redd construction and that the principle effect of cutting was to loosen the gravels and, hence, to increase porosity and the potential rate of intragravel flow. Grost *et al.* (1991) made comparisons of gravel composition in redd 'egg pockets', 'tailspill' and outside *Salmo* redds. The term 'egg pocket' appears to be used by these authors to describe the deposit of worked-over gravel in the downstream end of the pit, rather than the pockets within that deposit that contain concentrations of eggs (compare with the usage of the term 'egg pocket' in Chapter 2 and also by Chapman, 1988). If so, this is a similar comparison to that of Crisp & Carling (1989) but was made during winter as well as soon after redd completion. Redd construction gave statistically significant ($P < 0.10$ rather than the more usual $P < 0.05$) differences in the content of particle sizes of 3.4 mm diameter and less, though these differences diminished during the course of incubation. The amount of gravel cleaning that is achieved may well depend upon the amount of spawning activity taking place within a given area. *Salmo* spawning populations are often relatively sparse and this may lead to limited cleaning of gravels. *Oncorhynchus* species often spawn in very dense aggregations and redds may then be closely bunched with considerable overcutting. In such cicumstances substantial gravel cleaning is more likely. Kondolf *et al.* (1993) re-analysed data from 11 published studies and concluded that there was some cleaning of gravel during redd construction and that the final percentage of material with diameter < 1.0 mm (P_f) could be related to the initial percentage (P_i) by the equation

$$P_f = 0.63 \ P_i \hspace{4cm} (15)$$

4.2   Juveniles and smolts (parr for the watercourse)

Juveniles and smolts have similar general habitat requirements to those of sexually mature fish resident in fresh water. The swimming speed of salmonids and their ability to surmount obstacles are related to temperature. These relationships are therefore common to both freshwater residents and adults returning from the sea and are covered in the final section of this chapter.

4.2.1   Temperature

Simplified thermal tolerance polygons for the two *Salmo* species (Fig. 4.8) summarize much information on their temperature relationships. Comparison between the polygons for these two closely related species clearly demonstrates two important differences. First, the area enclosed in each tolerance zone is larger for the Atlantic salmon than for the trout. Second, the corresponding parts of the two figures are all at higher values (by between 2.0 and 5.0°C) for the salmon than for the trout, which implies that the trout is better adapted for low temperatures than the salmon. However, the two figures share several important features. For both species it is apparent that different acclimation temperatures can modify the upper lethal limits by several degrees but have only a very small effect (1°C or less) upon the upper limit for feeding. There is evidence (Spaas, 1960) that the upper lethal limit may also vary with the age/size of the fish. In contrast, the lower limit for feeding falls appreciably in fish acclimated to low temperatures and, given relatively gradual reduction in temperature, feeding can occur in both species down to 0°C. Examination of the figures suggests that the concept of a lower lethal limit is of little practical relevance in these two species. Exposure to temperatures below 0°C is highly improbable under natural conditions and it is clear that the lower lethal limit for both species falls above 0°C only for trout acclimated to temperatures of 18–25°C (or salmon acclimated

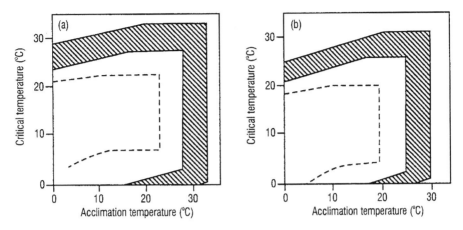

Fig. 4.8 Thermal tolerance polygons for *S. salar* (a) and *S. trutta* (b). The broken line indicates the limits for feeding. The shaded area is the critical range for survival. Its inner boundary is the incipient lethal level (survival time about seven days) and its outer boundary is the ultimate lethal level (death in ten minutes or less). The temperature limits for growth are indicated in Tables 4.3 and 4.4. Simplified from Elliott (1981, 1991).

to 16–32°C) that are then exposed to temperatures of about 3°C or less. It is difficult to imagine the occurrence of such circumstances in nature.

Inspection of the generalised salmonid thermal tolerance polygon (Fig. 3.7) shows that growth can occur only within a temperature zone that is smaller than the zone for feeding, that its upper and lower limits can be modified by acclimation, and that its lower limit is several degrees above freezing point. For *S. trutta* it is clear (Table 4.3) that the effects of acclimation temperature are relatively small. Acclimation can modify the upper limit for growth by about 1°C and the lower limit by about 1.5°C. It is probable that similar acclimation effects apply to *S. salar* though to rather higher upper and lower limits. More recent data (Elliott & Hurley, 1997) show that the upper and lower temperature limits for growth are all 2–3°C higher for *S. salar* than for *S. trutta* (Table 4.4). A number of publications (Elliott, 1975a, b, c; Elliott, 1976b, c) quantified the relationships between fish size, food consumption, ration size, growth and temperature for *S.*

Table 4.3 Lower and upper temperature limits for growth of *Salmo trutta* at two acclimation temperatures that are close to the actual limits. Derived from Elliott (1981).

Acclimation temperature (°C)	Lower limit for growth (°C)	Upper limit for growth (°C)
4.0	4.9	18.0
19.0	6.3	19.0

Table 4.4   Upper (T_U) and lower (T_L) temperature limits (°C) for growth and the optimum temperature value (T_M) for growth in brown trout and Atlantic salmon in fresh water. Derived from Elliott & Hurley (1997). These limits are for fish on maximum rations.

	S. trutta	*S. salar*
Upper temperature limit (T_U)	19.5	22.5
Optimum temperature (T_M)	13.1	15.9
Lower temperature limit (T_L)	3.6	6.0

trutta. Probably the most widely used relationship has been the one relating growth rate on maximum rations to temperature (Elliott, 1975a). Edwards *et al.* (1979) used this model, together with temperature data from 25 sites in UK streams and rivers, to show that the observed mean growth rates were between 60 and 90% of those predicted. This implies that natural food supply may be a less important constraint on growth than is often supposed but trout in some populations do not appear to conform to the model and show more rapid growth than predicted (Allen, 1985; Mann *et al.*, 1989; Crisp *et al.*, 1990; Jensen, 1990). These discrepancies may reflect acclimation effects, differences between populations, inadequacies of the temperature data, some combination of two or more of these factors, or that the model is faulty. In several streams in Wales, growth of brown trout appears to be only 30–70% of the maximum possible (Weatherley *et al.*, 1991), although this may simply reflect a difference in methodology. Most of the comparisons (*see above*) have been made by expressing the observed mean value of instantaneous growth rate (G) as a fraction or percentage of the predicted maximum value. In the Welsh study the observed weight at the end of a year's growth was expressed as a percentage of the predicted weight. This method gives a much larger percentage difference because of the compound interest nature of instantaneous growth rates. The growth model for trout has been updated recently and a similar model for *S. salar* has also been developed (Elliott & Hurley, 1997). The new model for trout gives predictions that are very close to those given by the earlier model but it is simpler than its predecessor and, as will be seen, all the growth parameters can be interpreted biologically. For both species the instantaneous rate of growth per day (G, as defined in Appendix B) can be predicted from the equation

$$G = cW^b \left[(T - T_{LIM}) (T_M - T_{LIM}) \right] \tag{16}$$

where W is fish weight in grams, T is temperature in °C, and c and b are constants. $T_{LIM} = T_L$ when $T \leq T_M$, and $T_{LIM} = T_U$ when $T > T_M$ (when T_U and T_L are the upper and lower temperature limits for growth, and T_M is the optimum temperature for growth. *See also* Table 4.4). The value of the constant b is 0.31 for both species; the value of c is 0.0280 for *S. trutta* and 0.0353 for *S. salar*. A plot of predicted values of G for salmon of four different weights (Fig. 4.9) shows how growth rate increases with temperature up to the optimum temperature for growth and then decreases as temperature rises above that optimum value. It is also apparent that at any given temperature the instantaneous growth rate decreases as fish weight increases. Comparison of plots for trout and salmon of 10 g weight (Fig. 4.10) again shows that the salmon has higher temperature values for its upper and lower growth limits and for its optimum growth temperature than does the trout. In addition, the instantaneous growth rate of salmon at its optimum temperature for growth is higher than that for the trout at its optimum temperature. As already noted, the values of upper, lower and optimum temperatures for growth can be modified by acclimation temperature and this should be borne in mind when applying predictive models that do not contain acclimation effects as an additional variable.

Inspection of published thermal tolerance polygons reveals, as already noted,

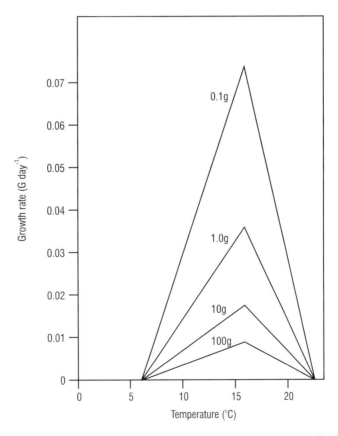

Fig. 4.9   Instantaneous rate of growth in weight/day (G) on maximum rations for *S. salar* parr of four different weights, relative to temperature. Calculated from equations of Elliott & Hurley (1997).

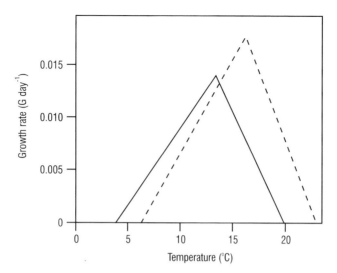

Fig. 4.10 Instantaneous rate of growth in weight/day^{-1} (G) on maximum rations for *S. trutta* (solid line) and *S. salar* (broken line) of 10 g weight, relative to temperature. Calculated from the equations of Elliott & Hurley (1997).

that both *Salmo* species are likely to feed at temperatures as low as 0°C. A number of studies suggest that at temperatures below approximately 7.0°C salmon parr tend to hide in the river bed, become relatively inactive and do not feed well (Allen, 1940; Saunders & Henderson, 1969; Gardiner & Geddes, 1980). Similar behaviour has been suggested for trout at temperatures below about 4.0°C. This may not be an adequate summary of the situation, for two reasons. First, data from Elliott (1981) show that the lower limit for normal feeding by *S. trutta* is about 4.0°C for fish acclimated to 20°C, about 3.0°C for fish acclimated at 10°C; and about 0.4°C for fish acclimated at 5.0°C. Second, recent evidence indicates that both *S. salar* parr (Fraser *et al.*, 1995) and small *S. trutta* (Heggenes *et al.*, 1993; Griffith & Smith, 1993; Vladimarsson & Metcalfe, 1998) may become predominantly night-active at low temperatures. It seems likely, therefore, that low temperatures may cause temporal changes in behaviour patterns (including feeding) rather than causing a total cessation of feeding. The switch to nocturnal activity at low temperatures becomes less marked with increasing size/age in *S. trutta* (Heggenes *et al.*, 1993; Pirhonen *et al.*, 1997).

In both species of *Salmo*, smolting and seaward migration depend upon the fish being physiologically ready to migrate before they respond to appropriate stimuli. Smolting appears to bear some relationship to fish size and Elson (1957) suggested that, as a general guide, *S. salar* parr normally smolt in the season following that in which they attain a length of 10 cm, though the ages of smolts may vary from one to six years (Metcalfe & Thorpe, 1990). Under good conditions, juveniles become smolts as I-group or II-group fish and the 'decision' whether or not to migrate as I-group appears to be made in July to August of the previous year. During the latter part of the 0-group year, a bimodal size distribution develops in the population. Members of the larger modal group become I-group smolts and members of smaller modal groups become II-group smolts (Metcalfe *et al.*, 1990). There is some doubt about the precise nature of the stimuli required to initiate a smolt run. Pemberton (1976a) observed that early runs of sea trout smolts to some sea lochs in Scotland were associated with heavy rain. Solomon (1978a, b) noted that the slightly increased turbidity and raised discharge after rain initiated major movements of salmon and sea trout smolts in chalk streams of southern England. These major movements were, however, mainly correlated with water temperature but the time of day of the movements was influenced by solar radiation. In each year the temperature value that initiated the movement was correlated with the mean daily water temperature during the 15 days before the start of the main run. Over a four-year period the 'switch temperature' varied from 10 to 12°C. In the River Thurso (Scotland) smolt migration was observed at temperatures as low as 7°C (Allen, 1944) and in the River Orkla (Norway) the smolt descent started at temperatures between 1.7 and 4.4°C (Hvidsten *et al.*, 1995). Thus, insofar as water temperature influences the timing of the smolt run, the 'trigger temperature' must be greatly influenced by acclimation temperature, must vary greatly between different populations or must be only one of a combination of factors operating. A number of authors have suggested that the major influences are water temperature and discharge (Forsythe, 1968; Österdahl, 1969; Baglinière, 1976). Hvidsten *et al.* (1995) analysed a complex array of factors and concluded that in the River Orkla the smolt run was influenced by water flow, water temperature, negative changes in water flow and temperature, phase of the moon, and also by interactions between smolts during the descent.

4.2.2   Water chemistry

The oxygen consumption per unit of body weight increases with temperature and with level of activity and decreases as fish size increases. Alabaster & Lloyd (1982) suggested that for salmonid waters, the annual 50-percentile of oxygen concentration should be at least $9 \, \text{mg} \, l^{-1}$, though the annual 5-percentile may be as low as $5 \, \text{mg} \, l^{-1}$. For adult salmonids, the 50- and 5-percentiles may be as low as $5 \, \text{mg} \, l^{-1}$ and $2 \, \text{mg} \, l^{-1}$, respectively, but these minima need to be increased for younger stages or when toxins are present. The sustained swimming speed of salmonids declines when dissolved oxygen concentration falls below 100% saturation, but swimming may continue to near-lethally low oxygen concentrations.

Inert suspended solids can have a variety of effects upon salmonid fishes. They may have indirect effects through reduction of light input and, when they settle out in slower flows, they may occlude gravel interstices and reduce the amount of hiding places for small fish and/or their invertebrate prey. More directly, they may abrade or clog delicate membranes (e.g fish gills) and they may cause skin irritation and abrasions, which may facilitate various secondary infections. There is no evidence of harm from concentrations of less than 25 mg l^{-1} and good or moderate salmonid fisheries can be maintained in waters with 25–80 mg l^{-1} (Alabaster & Lloyd, 1982). Concentrations of more than 80 mg l^{-1} do, however, occur for brief periods during natural spates in a number of rivers with good salmonid populations.

Bandt (1936) found that pH values of 9.2 and above were lethal to trout, though Alabaster & Lloyd stated that even values of 9.5 to 10.5 could be tolerated for brief periods. At pH values below 4.5 salt balance problems occur in trout (McWilliam, 1982). Values between 4.5 and 5.0 can be harmful, particularly in water containing low concentrations of calcium, sodium and chloride (Alabaster & Lloyd, 1982) and survival is reduced when these values are associated with elevated concentrations of aluminium (Brown, 1983). As already noted, pH and water hardness can interact to modify the toxicity of various metals and good examples are the combined effects of pH and aluminium on *S. trutta* (e.g. Sadler & Lynam, 1987) and the effects of pH and hardness on the toxicity of zinc to the same species (e.g. Everall *et al.*, 1989).

4.2.3 *Water velocity, depth and wetted area*

Three main types of study have given information on the effects of water velocity and depth upon the behaviour and distribution of juvenile trout and salmon. It is often difficult to distinguish the relative importances of depth and velocity. This is because they are often correlated with one another. Within a given portion of stream, water velocity tends to be less and depth greater in pools than in riffles, hence there is a negative correlation, spatially, between depth and velocity. At any given point in a stream there is, however, usually a positive correlation within a temporal framework in the sense that depth and velocity tend to increase and decrease together as stream discharge changes.

Some studies have been made in specially designed chambers or smooth channels to measure the ability of fish to withstand different water velocities. Heggenes & Traaen (1988) held very small *S. trutta* and *S. salar* in smooth channels and made step-wise increases in velocity to assess the 'critical velocity' at which each species was displaced downstream. The 'critical velocity' for swim-up fry was about 15 cm s^{-1} at 6–8°C and about 19 cm s^{-1} at 12–14°C, but after eight weeks feeding the velocity required to displace these fish downstream was over 50 cm s^{-1}. No clear difference in performance was found between *S. trutta* and *S. salar*. Shchurov & Shustov (1989) found that at any given size *S. trutta* were able to remain in the water column at their selected water velocity for longer than *S. salar*. At high velocities, however, salmon maintained station better because their large pectoral fins acted as hydrofoils. Such studies give an objective measure of the athleticism of the fish but do not take full account of any behavioural responses that the fish might show in a more natural habitat.

Studies in semi-natural channels can be used either for direct observation of the behaviour of the fish (Kalleberg, 1958) or for relatively precise measurement of the effects of water velocity (and depth) on their behaviour (Ottaway & Clarke, 1981; Irvine, 1986; Crisp, 1991; Crisp & Hurley, 1991a, b), though the design and practicalities of these experiments often give rise to some problems in interpreting the results. Kalleberg studied in detail the behaviour of juveniles of both *Salmo* species in a stream channel. He observed that salmon parr could hold station in surprisingly swift flows by adapting a 'hydrodynamically correct position' by use of the large pectoral fins. They also remained close to the stream bed, except at velocities below $10 \, \text{cm s}^{-1}$, when they moved to positions much higher in the water column. Irvine (1986) found that the numbers of recently emerged *O. tshawytscha* fry leaving a stream channel were greater during flow fluctuations than during steady flow. He had difficulty in relating his results to those of Ottaway & Clarke because their studies lacked a control in time. In *O. tshawystcha* and both species of *Salmo,* downstream movement occurs more readily by night than by day. In channels with natural gravel, nearly constant depth and four different velocities (measured at 0.6 of depth), the downstream dispersal of newly-emerged *S. trutta* was minimal at $25 \, \text{cm s}^{-1}$, rather higher at $7.5 \, \text{cm s}^{-1}$ and much higher at velocities above $25 \, \text{cm s}^{-1}$. In contrast, *S. salar* showed a very high rate of dispersal at $7.5 \, \text{cm s}^{-1}$ and much lower rates at higher velocities (25–$70 \, \text{cm s}^{-1}$). The final population density attained by both trout and salmon was not appreciably influenced by water velocity, except that at $7.5 \, \text{cm s}^{-1}$ the final density of salmon was very low. This suggests that the main influence of velocity was upon the rate of population density adjustment rather than upon final density, though salmon may actively avoid very low velocities.

There are a number of published field observations that describe the water depth and velocity preferences of trout and salmon. Analyses of the field distribution of underyearling *S. salar* showed maximum population densities where water velocity at $5 \, \text{cm}$ depth averaged 50–$65 \, \text{cm s}^{-1}$ and lower densities at velocities above and below this range (Symons & Heland, 1978). The densities of 0-group and I-group *S. salar* were inversely correlated with water depth and the densities of 0-group and I-group *S. trutta* were positively correlated with depth (Egglishaw & Shackley, 1982). Baglinière & Arribe-Moutounet (1985) found that 0-group Atlantic salmon preferred a stony substratum with high water velocity ($61 \, \text{cm s}^{-1}$) and reduced depth ($<23 \, \text{cm}$), whereas I-group and older trout preferred lower velocities ($<28 \, \text{cm s}^{-1}$) and greater depths ($>27 \, \text{cm}$). Heggenes & Borgstrøm (1991) noted that *S. salar* avoided areas with slow-flowing deep water and fine substrata but Gibson (1973) suggested that they would occupy low velocities in the absence of competitors. Food is obtained chiefly by the capture of drifting prey but also by foraging. For *S. salar,* at least, the abundance of drift (related to water velocity) is an important influence on its daytime distribution.

In both *Salmo* species changes in depth and velocity preferences occur as the fish grow in size. Bohlin (1978) found a negative correlation between numbers of 0-group and I-group trout at high population densities. Similarly, Symons & Heland (1978) found that 0-group (<7 cm) *S. salar* preferred 10–15 cm deep pebble riffles; larger specimens preferred deeper (30 cm) riffles with boulders; and yearlings (>10 cm) reduced the numbers of underyearlings (<6 cm) in these deeper areas by chasing them away. In the River Bush (Northern Ireland) over 70% of 0-group trout and Atlantic salmon were captured in mean depths <20 cm and older fish of both species were found mainly in mid-range depths. The data indicated that trout but not salmon were limited to areas of slower flow (Kennedy & Strange, 1982).

The observations described above refer to the situation during most of the growing season. There is accumulating evidence that with the coming of winter (and possibly in connection with the temperature effects noted in the previous section of this chapter) the depth and velocity preferences change for both species of *Salmo*. For example, Huntingford *et al.* (1988) showed that 0-group *S. salar* moved into areas of low water velocity during late autumn or early winter and that this was influenced by both date and water temperature.

There are several problems in the interpretation of these field observations. The first is through a lack of standardization between studies with regard to how, where and at what depth to measure water velocity. The second is that there is a lack of standardization in the definition of fish size, with some authors quoting ages, others quoting lengths, and some giving both. A final difficulty arises from the fact that because depth and velocity tend to be correlated, it is difficult to separate the effects of these two variables. For example, Egglishaw & Shackley (1982) made it clear that their results do not necessarily indicate a response by the fish to depth but rather to depth and/or some other factor(s) (e.g. velocity) correlated with depth. Almost the same point was made by Kennedy & Strange (1982). Nevertheless, a number of summary conclusions can be drawn from existing literature:

(1)   Through most of the year, juvenile *S. salar* prefer smaller depths and higher velocities than do *S. trutta* of similar size. Some studies suggest that young *S. salar* may actively avoid low velocities (<10 cm s^{-1}) but other observations suggest that they will occupy such velocities when not excluded from them by *S. trutta*.

(2)   It is difficult to quantify the precise velocity and depth preferences because of the difficulties of standardization that have already been mentioned. In general, juvenile *S. trutta* appear to prefer velocities of 25 cm s^{-1} or less, whilst *S. salar* prefer shallower water and velocities of 50–60 cm s^{-1} (but see points 3 and 4 below).

(3)   As they grow, the juveniles of both species of *Salmo* tend to move into deeper water.

(4)   In winter the behaviour patterns of *Salmo* change. The larger fish move into deeper water and lower velocities and spend more time under cover, whilst parr may shelter in the gravel and become more nocturnal.

(5)   Behaviour patterns and preferred habitat may vary between different times of day and night.

(6)   In channel experiments on newly emerged *S. trutta* and *S. salar* water velocity has an important influence on the rate at which population density is adjusted.

In addition to changes in water depth and velocity, changes in stream discharge also lead to changes in wetted area. Long-term change in mean wetted area will modify the area of stream bed available as habitat for salmonids and for their invertebrate prey. Rapid short-term fluctuations in wetted area may lead to the stranding of small salmonids.

Droughts, almost by definition, lead to long periods of low discharge and hence of shallow water, low water velocity, decreased wetted area, elevated temperature and, possibly, modified water chemistry. Therefore it is not surprising that relationships have been suggested between reduced survival of *S. trutta* fry (Elliott, 1986) and reduced growth of 0-group and older *S. trutta* (Weatherley *et al.*, 1991) and years of summer drought. The precise mechanisms of these effects are not clear and they may vary. Data collected from an isolated population of *S. trutta* in mid-Wales suggested that *S. trutta* recruitment may have been enhanced there during summers of drought, though this might reflect effects of water chemistry rather than direct effects of low flow (Crisp & Beaumont, 1998).

4.2.4   *Social interactions and habitat structure*

Social interactions in stream salmonids were reviewed by Gibson (1988). *Salmo* fry emerge from the gravel by night and rapidly disperse downstream from the redd site (Moore & Scott, 1988). Marty & Beall (1989) noted two waves of dispersal in *S. salar,* one soon after emergence and another 10 to 20 days later, at the time of assumption of territories. In some populations there is evidence of substantial downstream dispersal of 0-group trout during their first summer (Crisp, 1993b). As noted in Chapter 2, *S. salar* parr planted in spring usually seem to disperse less than 1 km, most for less than 100 m, during the first few months after swim-up.

There is evidence of both intra- and interspecific interactions in salmon and trout (Symons & Heland, 1978; Kennedy, 1982; Egglishaw & Shackley, 1982) but it is likely that the intensity of interspecific interactions may be reduced, to some extent, by the differing depth and velocity preferences of the two species.

The details of the physical structure of the habitat have an important influence on the choice and size of territories. Territory size decreases with visual isolation (Kalleberg, 1958) and this is influenced by water velocity affecting the depth at which the fish hold station and by the irregularity of the stream bed. The positions of territories are likely to be determined by physical factors such as food supply and the proximity of potential predators. The presence of predators and the occurrence of severe spates leads to the need within each territory for places in which to hide and/or take shelter. The precise type of hiding place is related to the size of the fish but, within this limitation of scale, suitable places can include irregularities of the stream bed, undercut banks, large rocks and boulders, fallen trees and branches. It is difficult to quantify the requirements for cover, particularly when its scale varies with fish size. What is clear, however, is that a stream in a relatively natural state with a wide variety of depths, velocities and bed materials and plenty of irregularities is likely to carry more salmonids than the more 'sanitized' watercourses that are favoured by many park-keepers and some civil engineers.

4.3 Sea life, adults and spawning (returning to the ancestral home and burying the evidence)

This section covers parts of the life cycle that, for anadromous salmonids, are spent partly at sea and partly in fresh water. In the previous two sections of this chapter, it was convenient to consider habitat requirements under the headings of individual environmental factors, even though this led to some overlap. Such an approach is not convenient for the present section, instead it is arranged under chronological headings.

4.3.1 Sea life

As the main concern of this book is with the freshwater part of the life cycle, this section on sea life will be relatively brief. Another reason for the brevity is that our knowledge of the marine part of the life cycle is much more limited than that of the freshwater phases. Attention will be concentrated mainly upon the genus *Salmo*.

The smolts of both *S. trutta* and *S. salar* usually move down river in spring and may reach the sea between April and July. They tend to aggregate in tight shoals before moving into the open sea (Pemberton, 1976a; Pratten & Shearer, 1983). In the past, the death of large numbers of smolts within heavily polluted estuaries (Meek, 1925; Pentelow *et al.*, 1933) was probably the main factor in the loss of the indigenous stocks of migratory salmonids from the major rivers of northeastern England. Sea trout generally remain within 80 km of their natal river (Berg & Berg, 1987) whereas salmon travel much further to various sub-Arctic feeding areas. The best known of these is around southwest Greenland, though one-sea-winter fish usually use feeding grounds closer to their natal river, for example around the Faroe Islands. Both species of *Salmo* show similar patterns in their return to their natal rivers. Individuals of both species may enter the river at any time of year. Those that enter fresh water in the spring remain in the river system and await the next spawning season (autumn/early winter). The returning adults often spend some time in the estuary awaiting suitable flows to enable them to move upstream and/or to acclimatize themselves to fresh water. A variety of factors is believed to induce migratory salmonids to enter the river system. Hayes (1953) made detailed studies on *S. salar* in a river in Nova Scotia (Canada). He concluded that natural freshets were capable of moving fish when other factors such as wind and tide were favourable but that major runs could also occur without freshets and could be maintained by a steady river flow. Temperature appeared to have little effect. Fish tended to move out of tidal waters at dusk and therefore light change may have been important. Hayes also noted that strong onshore winds induced salmon to concentrate in the estuary and eventually to ascend. More recent studies are in general agreement with Hayes' findings; they also reinforce the impression that this is a very complex subject. It is generally agreed that river flow is a major influence and that light level, temperature, tide and wind may also have some effect. It is also very likely that the effects of some or all of these variables may be interdependent (Smith & Smith, 1997). There are differences between rivers in the part of the tidal cycle that is associated with river entry (Hayes, 1953; Priede *et al.*, 1988; Potter, 1988; Webb, 1989; Potter *et al.*, 1992; Smith & Smith, 1997). One data set (Smith *et al.*, 1994) yielded adequate information for statistical analysis. The results showed that the degree of association between river entry and river discharge varied with season of the year. The absolute values of discharge were important in modifying the response to changes in discharge. The nature of the response to changing discharge was, itself, variable. During below-average (for the season) flows, there was a clear association between river entry and those days when flow had increased from that of the previous day. During higher than average flows, there was no such association. The authors concluded that models which relate river entry to absolute discharge values alone are not likely to be generally valuable. Smith & Smith (1997) found that in the Aberdeenshire Dee (Scotland), the upstream movement of *S. salar* occurred mainly at night and tended to coincide with the ebb tide. In contrast, in the River Usk (Wales), salmon migrated through the estuary on the flood tide (Aprahamian *et al.*, 1998). Smith & Smith stressed the practical difficulties that, even in these days of radio tags and electronic fish counters, make it difficult to draw reliable quantitative conclusions from such studies. In particular they refer to the disparity in findings from different studies and suggest that an important contributor to this may be differences in how and where salmon movements are monitored. It is also necessary to bear in mind possible effects of capture and tag-fitting upon subsequent behaviour patterns.

  Low dissolved oxygen concentration in river estuaries can be a problem and this can be aggravated by high water temperatures. Alabaster *et al.* (1991) showed that the lower lethal oxygen concentration for *S. salar* at temperatures between 15.0 and 27.5°C is given by the equation

$$\ln \ln C = 0.046\ T - 0.513 \tag{17}$$

where C is the lower lethal oxygen concentration as $mg\,l^{-1}$ and T is temperature in °C. In some polluted estuaries a plug of polluted water moves up and down with the tide and upstream migrants can pass only during very high river flows.

4.3.2   Upstream movement

Most studies have covered the requirements of migratory salmonids; resident trout have similar behaviour patterns and requirements during their movement to the spawning grounds, but allowance must be made for differences in fish size.

Depending upon the time of river entry and the length of the river, some fish will enter the river, move upstream, spawn and either die or return to the sea within a very brief timespan. Others spend nearly a year in the river and may move upstream in a series of stages between which they shelter in 'holding pools'. These resting places take the form of deep pools, boulders, undercut banks and fallen trees. The fish need them as refuges from predators and as protection from bright sunlight.

The stimuli for active upstream migration are difficult to quantify because there appear to be several of them and they probably interact with one another. Four can be identified:

(1)   Physiological readiness to spawn. This is probably the most important factor of all. Fish may move upstream months before spawning but it is likely that when the need to spawn becomes urgent then movement will occur regardless of the other factors. This is probably a major contributor to the difficulty of quantifying the importance of some of the other stimuli. Overripening of the eggs causes reduced viability in Atlantic salmon (De Gaudemar & Beall, 1998).

(2)   Temperature and dissolved oxygen concentration are both known to have some influence. Upstream movement of *S. salar* is inhibited at temperatures below 5°C (Pyefinch, 1955) and above 22°C and probably ceases altogether at a value between 22°C and 25°C (Alabaster, 1990; Alabaster *et al.*, 1991). Similar temperature criteria probably apply to *S. trutta* but with limits that are 2–3°C lower than those for *S. salar*. Upstream movement is an active process and relatively high dissolved oxygen concentrations are needed. The sustainable swimming speed of *S. salar* decreases when oxygen concentration falls to between 4 and $5\,mg\,l^{-1}$ (Beamish, 1978).

(3)   There is evidence that upstream movement occurs more readily during

darkness than during daylight and that water discoloration may serve as a substitute for darkness (Munro & Balmain, 1956; Hellawell *et al.*, 1974; Le Cren, 1985).

(4)   There is widespread consensus that river flow has an important influence upon the willingness of migratory salmonids to enter rivers and to move upstream. The relationships are difficult to quantify and more research is needed (Banks, 1969). There are two main areas of difficulty. First, most of the available data sets are inadequate because they lack information on temporal variation in the numbers of fish available to move. Further problems arise when daily counts are related to daily mean discharge without any allowance for the fact that there may be differences in fish movements between different parts of the hydrograph. Second, it is often assumed that the number of fish caught by anglers (or some variant upon the theme) is a good index of fish movement, but this is a dubious proposition (Hellawell, 1976). Similarly, it is often assumed that temporal variation in willingness to move will be reflected in the numbers of fish passing over a given weir or similar obstruction. In reality, what is probably being measured is the pattern of occurrence of conditions conducive to passage of that particular obstruction.

Additional potential problems include interactions between the different variables and also systematic temporal variation in some of them (particularly in physiological readiness to spawn). It is also possible that the stimuli needed to initiate upstream movement may differ in nature or degree from those needed to maintain it.

In the present state of knowledge we must conclude that most quantitative statements about the stimuli needed for upstream movement are empirical, based on studies in limited geographical areas and highly dependent on the conceptual model favoured by each author. Three types of model can be distinguished. In a study on a number of Scottish rivers (Baxter, 1961) it was assumed that adult salmonids required certain threshhold flows to be exceeded before they would move upstream. These flows were defined as percentages of the daily average flow (ADF) and the main details are summarised in Table 4.5.

Table 4.5   Summary of percentage values of average daily flow required before *S. trutta* and *S. salar* will move upstream in different parts of the river system. From Baxter (1961).

	Large spring salmon	Other salmon	Sea trout
Lower and middle reaches	50–70%	30–50%	20–25%
Upper reaches and headstreams	>70%	>70%	25–30%

Both Baxter (1961) and Le Cren (1985) noted that sea trout appear willing to move upstream over a wider range of flows than do salmon.

Several authors (Lamond, 1916; Huntsman, 1948; Stuart, 1957; Alabaster, 1970; Hellawell *et al.*, 1974) have suggested a rather more complex model in which the fish move only during certain parts of the hydrograph, usually the rising and falling limbs of the spate hydrograph rather than the spate peak. This type of model has been quantified for salmon in rivers in Lancashire and Cumbria (England) by Cragg-Hine (1985) on the basis of earlier work by Stewart (1969). In this model the flows required for movement are expressed as discharge per metre of river width. Salmon were inactive at flows below $0.03\,\mathrm{m}^3\,\mathrm{m}^{-1}$. Appreciable movement began at $0.008\,\mathrm{m}^3\,\mathrm{m}^{-1}$, peaked at $0.20\,\mathrm{m}^3\,\mathrm{m}^{-1}$ and decreased as flow increased further.

The third model postulates one or more peaks of movement that occur at certain times of year. Bimodal patterns have been observed in the River Coquet (Northumberland, England) by Banks (1969) and in the River Frome (Dorset, England) by Hellawell (1976). In the latter study there were peaks in June to August and in October to December. Unimodal patterns have been observed in the River Conon (Scotland) and in a fish pass with a constant, regulated discharge (Pyefinch & Mills, 1963; Gardner, 1971). Such seasonal patterns, apparently independent of direct physical stimuli, do not negate the other models but they could modify their practical implications.

Obstructions to upstream movement can arise from chemical barriers such as plugs of toxic or anoxic water, particularly in estuaries and the lower reaches of rivers. The most obvious obstructions are, however, physical barriers in the form of dams, weirs, rapids and waterfalls. The ease with which such obstacles can be passed may vary with discharge and water temperature. Some falls are surmounted by leaping and individual *S. salar* have been known to leap 3.7 m but the conditions for successful leaping are rather stringent. The fish usually leap from the 'standing wave' at the foot of the fall and the pool at the base of the fall should, ideally, have a depth of at least 1.25 times the height of the fall in order that the fish can gain sufficient swimming speed for the leap (Stuart, 1962). It is also probable that leaping height is related to maximum swimming burst speed (V_{max}) (*see* equation 19 below) and hence to fish size and water temperature. Rapids and flows over inclined surfaces on some weirs and falls are usually passed by high-speed swimming. The ability of salmonids to negotiate such obstacles depends upon the swimming capabilities of the fish in terms of both speed and endurance. In some instances a minimum depth of water on the inclined surface is also required.

The physiology of fish swimming is complex and there is a large body of literature. Swimming speed has been reviewed by Blaxter (1969) and Beamish (1978). The present account will be confined to a selection of information that is likely to be of practical value in river management. At least two different swimming speeds can be defined. The 'sustainable velocity' (V_{sust}) is the velocity that can be maintained over long periods without an oxygen debt being incurred. The 'maximum burst speed' (V_{max}) is the speed that can be attained in short bursts, but can be maintained only for a brief period. For *S. salar* a simple rule of thumb is the assumption that

$$V_{sust} = 2\ L\ s^{-1} \tag{18}$$

and

$$V_{max} = 10\,L\ s^{-1} \tag{19}$$

where L is fish body length in cm (Winstone *et al.*, 1985).

Bell (1986) recognised three speed categories:

(1) 'Cruising speed' can be maintained for extended periods and is $2–4\,L\ s^{-1}$. This corresponds to V_{sust} and is used during river migration.
(2) 'Sustained speed' ($4–7\,L\ s^{-1}$) can be maintained for several minutes and is used for passage of difficult areas.
(3) 'Darting speed' ($8–12\,L\ s^{-1}$) can be maintained for a few seconds and is used for escape and feeding.

Bell also gives summaries of ranges of velocities and indications of maximum leaping height for four species of *Oncorhynchus* and *Salmo trutta* (Fig. 4.11). As the performances indicated for *O. mykiss* are comparable with, or rather better than, those given for the three Pacific salmon species, we can assume that they include the anadromous form. The performance indicated for *S. trutta*, in contrast, is so poor that it must refer only to small specimens. A large sea trout can

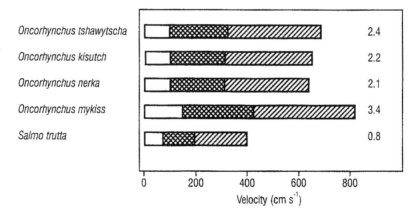

Fig. 4.11 Ranges of velocities ($cm\,s^{-1}$) for 'cruising' (plain bars), 'sustained' (cross-hatched bars) and 'darting' (shaded bars) swimming for *S. trutta* and four species of *Oncorhynchus*. The maximum leaping heights (m) are also shown. After Bell (1986).

certainly leap more than 0.8 m. Despite its value for general reference, this type of information is often of limited practical value compared with models that include fish size and water temperature as variables. The equation

$$V_{sust} = (0.32T + 8.0) \, L^{0.6} \tag{20}$$

is applicable to salmonids when T is temperature in °C and L is fish length in cm (Turnpenny, 1989). Computations of V_{sust} for fish of different lengths at five different temperatures (Fig. 4.12) show that the influence of temperature is considerable. Beach (1984) gives graphs of V_{max} for fish of various lengths at a series of temperatures. These can be used in conjunction with Turnpenny's equation to derive the relationship

$$V_{max} = V_{sust} \, (1.664 \, T^{0.2531}) \tag{21}$$

(Crisp, 1992). Plots of the calculated values of V_{max} (Fig. 4.13) show good general agreement, for fish of 40–75 cm length at 10–20°C, with the 'darting speeds' given for the three Pacific salmon species in Figure 4.11. The endurance time at V_{max}

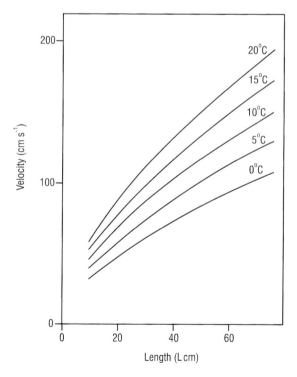

Fig. 4.12 Sustainable swimming speed for salmonids (V_{sust}, cm s^{-1}) relative to fish length (cm) at five different temperatures. Calculated from the equation of Turnpenny (1989).

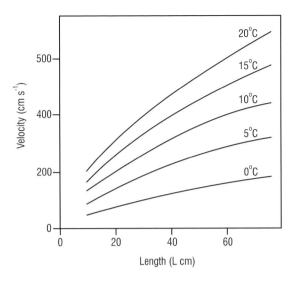

Fig. 4.13   Maximum swimming speed for salmonids (V_{max}, cm s^{-1}) relative to fish length (cm) at five different temperatures.

increases with fish size and decreases with temperature. The relationship is complex, but an approximate guide can be gained from Table 4.6.

This information on swimming speeds and endurance has potential value in river management because it can be used to assess the likelihood that a given barrier with known water velocities at given river flows will be passable by salmonids of a given size at known water temperatures. It is important to note, however, that predictions of V_{max} and of endurance must be taken as maxima for fish in peak physical condition. Injured, diseased, stressed and gravid fish are likely to have reduced capability.

4.3.3   Spawning

The time of spawning in salmonids is determined mainly by day length. This is a contrast with the Cyprinidae in which the time of ripening of the gonads is influenced predominantly by temperature (Bye, 1984). There are large differ-

Table 4.6   Approximate endurance times (s) at V_{max} for fish of four different lengths at four different temperatures. Approximate values derived from Beach (1984).

Temperature (°C) Fish length (cm)	2	5	10	20
10	5	2	–	–
25	70	25	12	3
50	900	220	70	40
75	–	1500	280	50

ences between rivers, within a salmonid species, in the timing and duration of the main spawning period. *S. trutta* spawns rather earlier than *S. salar*, though there is considerable overlap of the spawning periods of these two species in any given river. Within the UK in general the spawning of *S. trutta* is earlier and more synchronised in colder and more northerly rivers than in warmer rivers further south. This type of pattern is common to a number of species in a variety of regions, including *S. salar* in Norway (Heggberget, 1988) and *O. gorbuscha* and *O. tshawytscha* in Alaska (Sheridan, 1962; Burger *et al.*, 1985). It appears to reflect regional variations in incubation temperature and helps to ensure that swim-up coincides with rising temperatures in spring when food availability and the scope for growth will be increasing and, hence conditions for fry survival will be good (Heggberget, 1991). This probably reflects genetic differences between populations and appears to be a much more important adaptation than variations in rate of embryonic development.

Ready access for adult fish is a self-evident first requirement for a spawning site. Access to otherwise suitable sites may be prevented or impeded by natural or man-made obstructions and these may be either physical or chemical. The spawning site must have clean, well-oxygenated, running water and gravel of a suitable type. Some aspects of gravel composition have been covered already (*see* 4.1 Intragravel stages, and Chapters 2 and 3).

Published values of the water depths used for spawning vary from 15 cm to 90 cm (Fraser, 1975), with mean values of 38 cm for *S. salar* in Maine, USA (Beland *et al.*, 1982) and approximately 50 cm for both *Salmo* species in Norway (Heggberget, 1991). Most readers of this book will be familiar with television footage of *Oncorhynchus* spawning with their backs exposed above water. *Salmo* females may also behave in this manner but they usually choose water deep enough to cover them. There are few data sets that enable us to relate water depth and velocity to the sizes of female spawners though some information is available from the UK for the genus *Salmo*. A plot of mean water depth around each redd against female length (Fig. 4.14) shows that spawning occurred in a wide variety of depths. A calculated regression line relating the body depth of ripe female salmonids to their length is superimposed on this figure. It is clear that out of over 60 fish, only one spawned in less than her own depth of water. The maximum recorded depth was about 50 cm but fish were seen spawning in water much deeper than this. The upper limit of depth in this study was imposed by the fact that it is difficult to capture a female spawner for measurement in water deeper than 40–50 cm and the upper depth limit is, therefore, governed mainly by the length of the investigator's legs. Although trout and salmon will, if necessary, attempt to spawn in still water, they show a distinct preference for flowing water. Spawning in lakes usually occurs where there is upwelling flow from springs in the lake bed. Beland *et al.* (1982) quote water velocities for *S. salar* spawning sites in Maine (USA) of 25–90 cm s^{-1} at 12 cm above the river bed. Values from a Norwegian river (Heggberget, 1991) were lower with mean

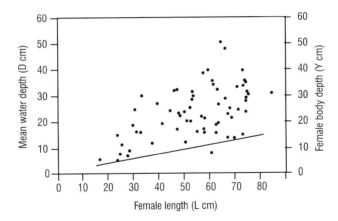

Fig. 4.14 Plots of mean water depth around the redd (D, cm) against female fish length (L, cm) for redds in northeastern England, southwestern Wales and southern English chalk streams. The data refer to both species of *Salmo*, but chiefly to *S. trutta*. The calculated regression line, Y = 0.176 L + 0.76, where Y (cm) is the body depth of ripe female salmonids, is also shown. After Crisp & Carling (1989).

values of 39 cm s^{-1} for salmon and 27 cm s^{-1} for trout. Fraser (1975) gave a range of velocities from 20–80 cm s^{-1}. The difficulty with such summaries is that it is always possible that the limits are imposed by circumstances rather than by the preferences or capabilities of the fish. Also it is likely that the limits will vary with the size of the fish. Beland *et al.* (1982) made the point that the salmon in their US rivers were of large average size (about 75 cm) and were able to spawn in higher velocities than would smaller fish. For *Salmo*, a plot of water velocity at 0.6 of depth at each redd site against female length (Fig. 4.15) takes the form of a 'wedge' of data points and the shape of this wedge can be described by means of two straight lines. The first is the line

$$V = 15 \tag{22}$$

where V is velocity in cm s^{-1}. This implies that all female spawners, regardless of size, tend to avoid velocities of less than about 15 cm s^{-1}. The second line has the equation

$$V = 2L \tag{23}$$

where L is fish length in cm. This implies an upper velocity limit that is related to fish size. Two points arise from this. First, the suggested lower limit of 15 cm s^{-1} is in reasonable agreement with the value of 20 cm s^{-1} given by Fraser (1975). Second, the line V = 2L falls close to, but a little below, the predicted sustainable swimming speeds at 5°C (*see also* Fig. 4.12) which is the likely water temperature at the time of spawning in the UK (*see* means for October and November for

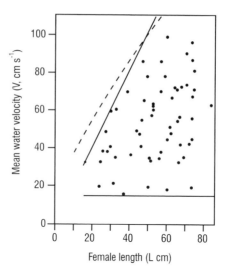

Fig. 4.15 Plots of mean water velocity (V, cm s^{-1}) at 0.6 of depth around the redds against female fish length (L, cm). The lines V = 15 and V = 2L are shown as solid lines and the predicted sustainable swimming speeds at 5°C are shown as a broken line. Other details as in Figure 4.14.

northern streams in Table 3.5, and for November, December and January for southern streams in Table 3.6).

There is a general tendency for sea trout to spawn in higher headwaters, whilst most salmon spawn further downstream. It is also apparent that the upstream limits for sea trout may be lower in dry than in wet spawning seasons.

There have been many investigations of the composition of spawning gravels and there is general agreement as to the ideal composition. From a detailed study of *Salmo* redds, Fluskey (1989) concluded that a suitable mixture for use in stream rehabilitation or in artificial spawning beds was 10% 'cobbles' (64–190 mm), 35% very coarse gravel (32–64 mm), 25% coarse gravel (16–32 mm), 20% medium gravel (8–16 mm) and 10% fine gravel (4–8 mm). We might also expect that the size of gravel in which a given female could create a redd might, amongst other factors such as water velocity, be related to her size. A study of a large number of redd sites of two species of *Salmo*, six species of *Oncorhynchus* and one species of *Salvelinus* (Kondolf & Wolman, 1993) showed a relationship between female fish length and the mean value of the median grain size of the gravel used (Fig. 4.16). It was also possible to fit, by eye, an approximate relationship between female fish size and the maximum value of median grain size likely to be used for spawning.

Most of the published studies on the hydraulics of habitat utilisation for spawning either fail to consider habitat availability or tend to treat it as a static variable. Moir *et al.* (1998) drew attention to this problem and noted the need to

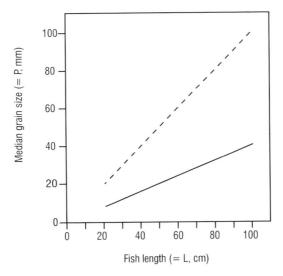

Fig. 4.16   Relationships between female fish lengths (L, cm) and the median grain sizes (P, mm) of gravels used for spawning by salmonids. The solid line is the mean, with equation P = 0.4 L. The broken line is an approximate indication of the maximum possible values and has the equation P = L. After Kondolf & Wolman (1993).

consider habitat utilisation within the framework of habitat availability which is, in turn, a function of stream discharge.

It is now appropriate to consider briefly the influence of the parents, especially the female parent, upon the probability of survival of their progeny. The choice of redd site is of crucial importance because, at least up to the time of hatching, the embryos are unable to move to more suitable places in the event of the redd becoming unsuitable for successful incubation. Another important aspect is the size of the female fish. It is often suggested that large size confers some advantage to a female fish and/or to her progeny (e.g. Holtby & Healey, 1986, for *O. kisutch*). In general, within a natural population of a salmonid species, the larger females produce larger eggs. These larger eggs tend to produce larger fry and large fry size is believed to give advantages in competition for territories and also to enable the fry to survive longer when starved (Elliott, 1984a). It is logical to assume that as larger females tend to bury their eggs more deeply than do smaller females, the eggs of the former will be less prone to washout during spates. There is, however, also the possibility that deeper egg burial will increase the risk of death through asphyxiation or entrapment, as a result of the deposition of fines during the incubation period. Holtby & Healey (1986) could find no evidence that the eggs of larger females consistently survived better than those of smaller ones, though egg mortality appeared to be influenced by gravel quality and bed scour and one would expect losses through bed scour to be related to burial depth. A comparison between female *S. trutta* that migrated downstream into a reservoir and females that remained in the afferent streams showed that, on average, a stream resident produced more eggs in her lifetime than a female that entered the reservoir and attained large size (Crisp, 1994). As most females in this population did, in fact, migrate to the reservoir, it is probable that the advantage of large female size was related to the larger eggs produced and/or to the greater depth at which they were buried.

Chapter 5
Human Influence

('We plough the fields and scatter...'
– slurry by the burn?)

Summary

Human impacts upon salmonid rivers must be considered in terms of land use within whole catchments. Small tributaries are important salmonid spawning and nursery areas and the health of drains and streams is essential to the well-being of the whole river system. Civil engineering works can damage salmonid habitat during the construction phase via inputs of sediment, spillage of harmful materials and damage to gravel beds. In addition the completed structures can themselves have harmful impacts. Examples are habitat degradation caused by drainage, stream training and flood protection works; obstruction of fish movement by weirs and dams; modification by impoundments of downstream flow, silt and gravel movement regimes, and of water temperature and chemistry patterns; damage to smolts by hydroelectric installations; and over-abstraction of ground and surface waters. Sound and rational management of fishing and fisheries is theoretically possible, but really requires a more extensive science base before it can be fully realized. Fish farming, both in fresh waters and the sea, has been a growth industry in recent decades. It can cause problems in water quality, the proliferation and spread of diseases, and genetic damage to natural populations as a result of inter-breeding with farmed fish. Past and present industry can pose threats to the aquatic environment, chiefly through various types of pollution. Road construction and road transport can lead to large inputs of suspended solids and of some chemical pollutants to watercourses. Problems from urban development arise chiefly from water consumption, sewage production, waste production, atmospheric pollution and miscellaneous accidental pollution incidents. Forestry impacts relate chiefly to modifications to run-off quantities and patterns, suspended solids inputs and acidification. The nature and magnitude of the various effects varies with soil type, stage in the forestry cycle and management regime. Recent publications have helped to quantify some of the effects of forestry but the precise mechanisms are still not very clear. Agriculture has a wide variety of effects and these vary between localities and between individual farms in their nature and intensity. Major issues are suspended solids inputs chiefly from trampling, overgrazing and damage to river banks; increased biochemical oxygen demand (BOD) caused by run-off of slurry and silage effluents; and pollution of streams by sheep dip. It is difficult to quantify agricultural effects because they are widespread, diffuse and spatially very variable.

Both the qualitative and the quantitative aspects of human influence upon salmonid fishes will vary considerably both within and between different countries and regions. Nevertheless, many of the general principles are likely to be common to most regions. Therefore the main quantitative statements made in this account refer either to Great Britain (Scotland, Wales, England and various associated islands) or to the United Kingdom (Great Britain together with Northern Ireland).

Within Great Britain, nearly half of the total land area is tilled land and managed grassland, and 28% is rough grass, marsh and moorland (Fig. 5.1). It is

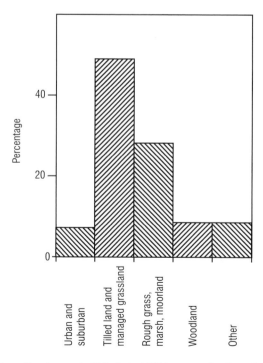

Fig. 5.1 Distribution of land cover in Britain in 1990 between five broad categories, expressed as percentages of the estimated total of 240.3 thousand km^2, derived from Royal Commission on Environmental Pollution (1996).

important to recognize that a river system is greatly influenced not only by activities within the main river and on its adjacent banks, but also within the whole catchment area. The 'health' of the main river, in terms of its water quality, flora and fauna, is heavily dependent upon the health of its numerous tributaries of all sizes. Estimates of the relative proportions of main stream and tributary within a given system are usually based on the study of maps. Therefore the results obtained are heavily dependent upon the assumptions underlying the methodology and upon the scale of map used. Smith & Lyle (1979) estimated the number of 'rivers' in the UK as 10083 and the number of associated 'streams' as 194674. On a much smaller scale, Trout Beck is a tributary of the upper River Tees and a map and data published by Crisp *et al.* (1975) can be updated to show that, using a map of 1:63630 scale, this stream has a total area of about $27455\,m^2$, of which about $13125\,m^2$ (3750 m of stream length) are main stream and about $14330\,m^2$ (17070 m of stream length) are tributaries. At least 75% of the total length of these tributaries is large enough to be useful trout habitat. Whatever the methodology used, it is clear that tributaries form a large component of each river system and in salmonid rivers they are particularly important as spawning and nursery areas. If the environmental quality of small drainage ditches is neglected, then environmental quality throughout the whole system is likely to suffer.

Figure 5.2 lists a number of ways in which humans can intervene in the aquatic environment and the main possible consequent effects upon trout and salmon, either directly or via their environment. These various interventions will be described in more detail below. Three points should, however, be borne in mind.

(1) Some human activities may include several different interventions. For example, agriculture may include water abstraction, land drainage, channel dredging, farm effluent, extraction of sand and gravel, changes in bank vegetation, changes in stream vegetation, roads and vehicles.
(2) Not all of the listed consequences will necessarily follow from any individual act of intervention. For example, there is great variation between different industrial effluents and any given effluent might have any, all or none of the consequences listed in the table.
(3) 'Changes' refer to both qualitative and quantitative modifications. For example, a change in aquatic vegetation may take the form of a change in the species composition of the flora or in its biomass, or in both. Similarly, a change in the bank vegetation may take the form either of a change in the type of crop being grown or of a change in the husbandry (e.g. fertilizer or pesticide regime) of an existing crop.

In the following account we shall consider human influences under several headings that reflect major fields of human activity, though appreciable overlap can occur between categories. At various points there will be reference to 'bio-chemical oxygen demand' (BOD). When water contains organic matter that is

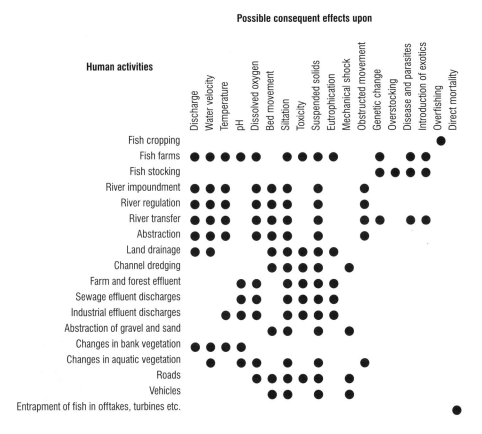

Fig. 5.2 List of human activities and some of their possible consequent effects on trout and salmon. After Crisp (1993b).

being broken down by bacteria, the process of decomposition consumes oxygen. The BOD of a sample is a standard measure of the amount of oxygen per litre of sample that will be consumed over a five day period at 20°C in darkness, given a suitable innoculum of bacteria to start the process.

5.1 Civil engineering, with particular reference to water storage and supply

This section deals with some generalities of civil engineering activity and, in a little more detail, with various aspects of water supply. Some other aspects are covered in later sections on transport and urban development.

Civil engineering effects on the aquatic environment contain two main elements. First, there are the impacts of the actual construction work; second, there are those effects arising from the existence and operation of the completed project.

Construction work includes such schemes as flood protection works, stream channel alteration, river crossings for roads or pipelines, and instream structures, including weirs, flumes and reservoirs. Major problems for salmonid fishes during construction work usually arise from habitat disturbance or degradation, inputs of large amounts of sediment (Fig. 5.3) and accidental spillages of diesel fuel and similar noxious materials. Habitat degradation arises from the removal or disturbance of gravel beds used by spawning fish or through simplification/sanitation of the stream structure leading to reduction of available fish habitat. The mobilization of large amounts of suspended solids may result in the infilling of gravel pores to the detriment, particularly, of intragravel stages. The passage of machines over gravel beds during the incubation period may further influence the survival of intragravel stages (Fig. 5.4).

Fig. 5.3   The confluence of two streams, showing one (bottom) laden with suspended solids from civil engineering activities upstream. England, *c.* 1988. Photograph by Prof. Paul Carling.

Engineers and others have often carried out river training and flood protection schemes with scant regard for the hydraulics of the system or for the physical or biological consequences (Fig. 5.5). Some of the most obvious examples of this arise when small flashy streams are straightened by farmers or contractors. This usually leads to increased instability of bed and banks and to increased erosion, often to the detriment of the perpetrator of the 'improvements', or to others. In recent years there has, however, been increasing recognition of such problems and of the need to carry out any necessary works in a more sympathetic manner (e.g. Vivash, 1989).

Such completed structures as flumes, weirs, dams and barrages have the

Fig. 5.4 An agricultural tractor passing over a gravel bed. Wales, *c.* 1985. At the wrong time and in the wrong place, such activity can damage the intragravel stages of salmonids. Photograph by Prof. Paul Carling.

Fig. 5.5 A heavily canalised stream course that will have limited value as salmonid habitat. England, 1997. Photograph by Prof. Malcolm Newson.

potential to interfere with salmonid migration and the need for provision of effective fish passes or of bypass streams is self-evident. In addition, large impoundments (chiefly for water supply) can modify the flow regime, water temperature and water chemistry downstream.

The flow regime of a natural river is influenced by the geology and soils, relief, climate, vegetation cover and land-use of its catchment. Human intervention through the use of reservoirs and between-rivers transfer schemes can modify the flow regimes of rivers (Fig. 5.6) by increasing or decreasing total discharge or by modifying the pattern of discharge fluctuation.

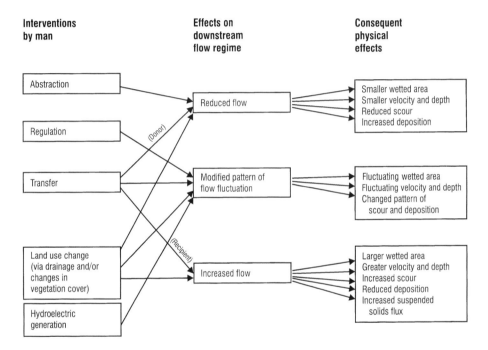

Fig. 5.6 Simplified summary of major modifications to the flow regime that can occur as a result of human intervention in the system. Some of the main physical consequences that can result from such intervention are also indicated. Redrawn from Crisp (1993b).

Reservoirs for the supply of domestic and industrial water are of two main types. A 'direct supply reservoir' supplies water via pipelines directly to the treatment plant. This leads to a reduction in discharge to the downstream river and to substantial reduction of discharge fluctuations. In contrast, a 'regulating reservoir' stores water at times of high natural river discharge (usually winter and spring) for release at times of low natural river discharge (summer and autumn). This ensures adequate flow at some downstream point(s) for abstraction of water for treatment and supply. A regulating reservoir therefore modifies the temporal pattern of discharge downstream but does not reduce the total downstream discharge (apart from any losses arising from evaporation during reservoir storage). There will, however, be a reduction of discharge at the downstream abstraction points. In common with direct supply reservoirs, regulating reservoirs cause a marked reduction in short term discharge fluctuations. This is clearly seen (Fig. 5.7) for the River Tees below Cow Green regulation reservoir (northern England) where there has been an almost complete elimination of discharges less than one tenth and greater than seven times the mean discharge. In the natural river, discharges less than one tenth of the mean value occurred for about 20% of the time and and there were substantial spates. An inter-river transfer of water leads to reduced discharge in the donor river and increased discharge in the recipient river, and to modified patterns of discharge fluctuation in both. Such schemes may also result in transfers of pollutants, parasites, diseases and fish species, modification of the genetics of fish in the recipient river, and disruption of the homing mechanisms of migratory salmonids.

The feature common to impoundment, transfer and hydroelectric schemes is that they modify discharge in terms of the quantity of water and/or the patterns of discharge fluctuation. The consequences of this are changes or fluctuations in

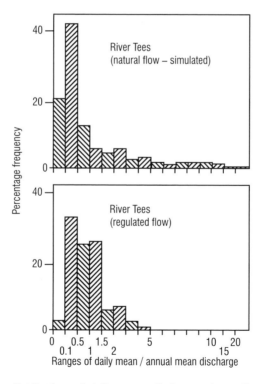

Fig. 5.7 Frequency distribution of daily mean discharges from Cow Green Reservoir expressed as proportions of the mean annual discharge, in the natural River Tees (computer simulation) and in the regulated River Tees (observed values). After Crisp (1984).

wetted area, water depth, water velocity, bed scour and fine sediment transport and deposition. As we have already seen, changes in wetted area may modify the area available for spawning or juvenile habitat and rapid changes may lead to exposure of eggs to desiccation or freezing and to the stranding of small fish. Somme (1954, 1960) estimated that 35% of the sea trout eggs laid downstream of an impoundment on the Eira River (Norway) were destroyed through exposure to freezing during the 1953–1954 incubation period. This may, however, reflect particularly severe conditions. In most circumstances the moisture held in the gravel interstices and the thermal insulation provided by the gravel may sustain eggs for several hours or even weeks (Hardy, 1963; Hawkes, 1978).

Changes in water velocity and depth may modify the availability of spawning sites for salmonids and, via changes in intragravel flow, may influence the survival of intragravel stages. The relative favourability of stream habitat to juveniles of different species may also be modified. Changes in the amount and location of bed scour and sediment deposition may influence the survival of intragravel stages and the amount of cover available for juveniles. At some impoundments it is possible to make releases that equal or exceed the discharge of natural spates. At such structures it is important to ensure that these large releases, either during commissioning trials or during routine operations, do not wash out spawning gravels. Should this happen in the presence of the dam and the absence of tributaries, there will be no source of gravel to replenish the deposits and useful spawning/nursery areas may be lost. Such events have been recorded, but as most impoundments in the UK tend to be made in upland areas on spate rivers, the inverse problem is much more common. It is usually not possible to make releases that have even approached the spate discharges in the natural river (Fig. 5.5) and, over a period, the gravels between the dam and the nearest major downstream tributary may become compacted, infilled with fines and of considerably reduced value as a spawning medium.

The general effects of impoundments on water temperature at the point of release are similar for nearly all reservoirs but differ quantitatively according to reservoir size. We will consider the general principles in terms of two English reservoirs and then briefly set them in a wider context. Cow Green Reservoir in County Durham is of average size by European standards (312 ha). It is relatively shallow (23 m), in an exposed position and stratifies only rarely and briefly. Kielder Water in Northumberland is large size (1087 ha) by European standards, but small in a world context. It has a maximum depth of 52 m and it undergoes stable stratification each summer. When the annual temperature cycle of the water discharged from each reservoir is compared with the cycle in the natural river (Fig. 5.8; Table 5.1) it is apparent that the effects of impoundment are common to both reservoirs and comprise:

(1)   a later rise in water temperature in the spring;
(2)   a later fall in water temperature in the autumn;

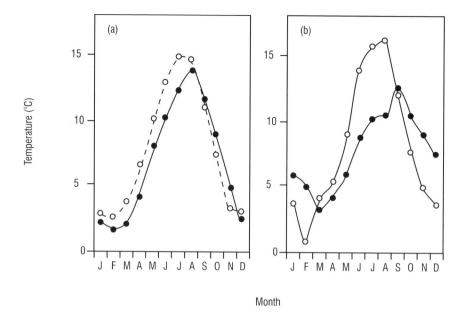

Fig. 5.8 Monthly mean water temperatures for large natural tributaries (solid lines and filled circles) and for the regulated river (dotted lines and open circles) at Cow Green Reservoir (a) and Kielder Water (b). Redrawn from Crisp (1985).

(3)   a lower summer peak;
(4)   reduced amplitude of monthly means;
(5)   reduced diel range.

There are also some striking differences between the two downstream temperature regimes. These may, in part, relate to the fivefold differences in reservoir capacity (Table 5.1) but probably reflect mainly the presence of summer stratification at Kielder but not at Cow Green. Thus the downstream temperature effects at Cow Green arise from storage effects, whereas the effects at Kielder arise from the combined effects of storage and stratification. Two important differences are the much greater delay of the annual temperature cycle and greater depression of the summer peak at Kielder than at Cow Green. In addition, impoundment at Kielder has caused substantial elevation of winter temperatures but there is no evidence of this at Cow Green (if anything, the reverse is shown). It is also worth noting that the temperature of water released at Kielder during stratification is largely a consequence of the choice of draw-off levels made by the engineers. It is clear from Fig. 5.9 that during July and August suitable choice of draw-off level could give release water temperatures ranging between about 7°C and 18°C.

We have, so far, considered the temperature effects at the point of release and, hence, at what is likely to be the point of greatest biological impact. As the water

Table 5.1 Summary of the temperature regimes (n) in the natural river and (i) immediately downstream of each of two regulating reservoirs in northern England. *Mean values 2.1°C during the period of stratification and 0.8°C during the rest of the year. Derived from Crisp (1985).

	Cow Green Reservoir		Kielder Water	
Area (ha)	312		1087	
Capacity (m³ × 10⁶)	41		201	
	n	i	n	i
Annual mean (°C)	7.8	6.8	8.2	7.8
Amplitude of monthly means	12.4	12.1	15.4	9.3
Mean daily range (°C)	4.2	0.6	2.6	1.1*
Summer peak (°C)	15.0	13.7	16.3	12.6
Time of summer peak	July	August	July/August	September
Winter low (°C)	2.6	1.6	0.9	3.3
Time of winter low	February	February	February	March

passes downstream its temperature will move towards that of the air and the entry of unregulated tributaries will further reduce the effects of impoundment. There are few good data sets that enable us to predict the rate of reduction of these temperature effects with distance downstream. The general consensus is that for reservoirs of the sizes usually found in Europe (several tens to several hundreds of m³ × 10⁶) the effects reduce rapidly with distance downstream and are generally undetectable or barely detectable at 30 km, or perhaps less, downstream of the release point (Table 5.2). In a world context, a number of

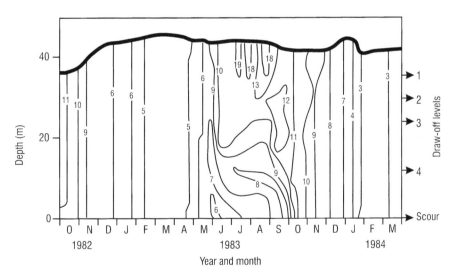

Fig. 5.9 Depth–time diagram for water temperatures in Kielder Water (northern England). The draw-off levels are also indicated. The thick line represents the water surface. The thinner lines are isotherms (lines of equal temperature) and the numbers on the body of the graph are °C. Redrawn from Crisp (1985).

Table 5.2 Summary of the distances downstream at which the temperature effects of impoundment can be detected for reservoirs of different sizes. Derived from Crisp (1985).

Reservoir	Source of information	Capacity ($m^3 \times 10^6$)	Downstream effects on water temperature
Vir River Valley (Czechoslovakia)	Penaz *et al.* (1968)	53	Effects substantial 7 km downstream; barely detectable at 30 km.
Cheeseman Lake (Colorado, USA)	Ward (1976)	98	Marked effects at release point; no discernible effect 32 km downstream.
Lake Hume (Murray River, Australia)	Walker *et al.* (1979)	3070	Effects still apparent 200 km downstream.
Fort Randall (South Dakota, USA)	Neel (1963)	7524	Effects detectable at least 130 km downstream.

reservoirs have capacities measured in thousands of $m^3 \times 10^6$. These have much larger effects on the temperature of the released water and this, together with the much larger size (hence, thermal capacity) of the rivers that they feed, results in the temperature effects being measurable and probably biologically important for hundreds of km downstream (Table 5.2).

The importance of temperature in the biology of salmonid fishes was stressed in Chapter 4. It is now possible to use some of the predictive models to assess some of the probable effects on salmonids of the changes in temperature regime measured at three different reservoir systems. Cow Green and Kielder reservoirs are already familiar. The River Elan (Wales) is impounded by several reservoirs that have a combined capacity of $51\,m^3 \times 10^6$ and an area of 340 ha. The Elan reservoirs stratify and their effects on downstream temperatures are, despite their smaller size, similar to those of Kielder apart from a much smaller elevation of winter temperatures. For each example, the annual temperature cycles for the natural and the regulated river have been used to predict the annual maintenance ration (the ration required to keep the fish alive without either increase or decrease in weight) for an 18.6 g trout and the mean instantaneous growth rate per day (*see* Appendix B) through the year for a trout whose weight was 18.6 g on 1 January. The percentage changes in the predicted values as a result of impoundment are shown in Table 5.3. The smaller amplitude of the annual temperature cycle under regulation leads to a reduction in the predicted maintenance ration at all three sites, ranging from 14% at Cow Green to 20% in the Elan. Predictions of mean instantaneous growth rate show very different effects at the three sites. At Cow Green, despite changes in the annual temperature cycle, the increases and decreases in growth rate at different times of the year balance one another to give a prediction of no change in mean growth rate on maximum rations. This prediction is confirmed by field studies in which observed

Table 5.3 Percentage change in predicted annual maintenance ration for a brown trout with a constant weight of 18.6 g and in predicted mean instantaneous growth rate per day on maximum rations for a trout that had a weight of 18.6 g on 1 January, as a result of temperature changes caused by impoundment. Calculated from equations of Elliott (1975a, b). This refers to the region immediately downstream of the impoundment.

	Annual maintenance ration (g dry weight)	Mean instantaneous growth rate (G) per day on maximum ration
Cow Green	−13.9	0
Kielder	−17.5	+18.2
Elan Valley	−19.8	−12.2

growth was at 80% of the predicted rate both before and after impoundment (Crisp, 1977). At Kielder, the effect of impoundment was to bring temperatures in both summer and winter closer to the optimum for growth and this gives an 18% increase in the predicted growth rate. In contrast, in the River Elan, a decrease in growth rate of 12% is predicted.

It is clear from these results that the effects upon growth rate of the downstream water temperature changes caused by impoundment are likely to vary considerably between sites dependent upon the fine details of the relationship between the annual temperature cycle and the optimum growth temperature for the fish species concerned. As *S. trutta* and *S. salar* have different optima and upper and lower limits for growth (Fig. 4.10; Table 4.4), the effects of any given impoundment upon growth may vary between the two species and possibly between years. The spawning time of the *Salmo* species in any given stream is determined mainly by day length with temperature having very little effect (Bye, 1984). Therefore impoundment is unlikely to modify the spawning time of salmonids in the downstream river. It will, however, modify downstream temperatures and hence the rate of intragravel development. Water temperatures for the natural and regulated River North Tyne at Kielder have been used to predict the development of the intragravel stages of *S. trutta* and *S. salar*, given an oviposition date of 15 November, immediately downstream of the release point. For both species the predicted date of median swim-up was several weeks earlier under regulation than under natural temperatures. Before impoundment, swim-up in *S. trutta* would have occurred about two months after the temperature rose above 4°C and *S. salar* about one month after 7°C was attained. After impoundment, *S. trutta* would swim up about 13 days before 4°C was attained and *S. salar* about 50 days before the attainment of 7°C. Assuming that at temperatures below 4°C for *S. trutta* and 7°C for *S. salar* the fish will not feed and will therefore starve, then the effect of impoundment would be to impose upon both species the possibility of starving to death soon after swim-up. The probability of surviving would be higher for *S. trutta* than for *S. salar*. This exercise is an oversimplification in two respects. First, it assumes fixed lower limits for feeding for both species, but in reality the values will vary with acclimation temperature (Fig. 4.8). Second, despite earlier beliefs, these supposed lower limits for feeding may actually be the temperatures at which feeding is switched from daytime to the hours of darkness. Nevertheless, it does show that the temperature changes immediately downstream of the release point will cause some stress to the fry of both species but will disadvantage *S. trutta* less than *S. salar*.

Some impoundments may modify the flux of ions in solution in the downstream river, especially during the first few years after construction when nutrients may be released by the decomposition of inundated vegetation. The main lasting effect, however, is to modify the temporal pattern of fluctuation in concentration. At Cow Green Reservoir there was no clear evidence of any change in annual ionic flux but the pattern of fluctuation changed from one in which ionic concentrations were mainly related to discharge to one in which the fluctuations were smoother and had a seasonal basis. There is little evidence that these changes affect salmonid fishes.

When hypolimnetic water is released from reservoirs there may be some deoxygenation in downstream areas. In typical shallow, fast-flowing and turbulent upland rivers (at least on a European scale), reoxygenation will be rapid and extensive deoxygenated areas are unlikely (Edwards & Crisp, 1982).

Reservoirs act as settling ponds and generally receive considerably more inert suspended solids than they release. In examples where there is a substantial oxygen deficit however, there may be local accumulation of a dark brown deposit consisting of about 40% organic matter, 6% iron and 4% manganese. In the River Elan downstream of the reservoirs, this material can accumulate at $0.1–0.2\,\mathrm{g\,m^2\,day^{-1}}$ (Edwards & Crisp, 1982). Its effects upon salmonid fishes have not been studied in detail but it is likely to have a harmful effect on spawning gravels and on invertebrate food organisms in the downstream river.

Hydroelectric impoundments usually have little effect on the total discharge but they can create highly unnatural discharge fluctuations in the river downstream and such fluctuations often bear a closer relationship to the demands of the National Electricity Grid than to the needs of the river biota. Two additional problems are associated with hydroelectric schemes. The first is that salmonids, especially downstream migrating smolts, can be damaged by passage through turbines. Struthers (1989) distinguished between low-head ($<30\,\mathrm{m}$) and high-head ($>30\,\mathrm{m}$) turbines. He commented that smolts usually pass through low-head turbines with negligible loss but that high-head turbines can cause appreciable mortality. Estimated mortality rates at Scottish high head turbines vary from 20 to 60% at Invergarry (Munro, 1965) to 3 to 15% at Laing Dam (Struthers, 1989). At most sites, efforts have been made to prevent smolts being drawn through turbines and to provide a safe alternative route to the downstream river. There is a wide variety of techniques to exclude fish from turbine intakes (Turnpenny, 1989) but Struthers (1989) advised that at some sites removal of the screens might be beneficial 'on the basis that the mortality rate through the turbines is much less than that caused by failure of the smolts to find the safe bypass route and the associated predation resulting from migratory delay'. Secondly, in certain circumstances the water passing over the spillways or through the turbines of very high hydroelectric dams can become supersaturated with nitrogen. This leads to supersaturation in the body fluids of fish in the downstream river and to subsequent formation of gas bubbles in the tissues of the fish. This is known as 'gas bubble disease' and is related to the disease known as 'the bends' that affects human divers who are decompressed too rapidly. At some sites in the USA dissolved nitrogen concentrations of up to 145% saturation have been observed and these can be accompanied by substantial mortalities of both juvenile and adult salmonids (Beiningen & Ebel, 1970; Raymond, 1979). The severity of the effects can be related to the level of supersaturation, duration of exposure, water temperature, the general physical condition of the fish, and the swimming depth maintained by the fish (Raymond, 1979). Many Scottish hydroelectric dams are over 100 m high; the highest is over 400 m. Supersaturation could occur at some of them. Despite this, gas bubble disease has not been identified as a problem at any Scottish scheme (Struthers, 1989).

A problem that is, perhaps, most severe in the south and east of Britain is overabstraction from both ground and surface waters. This can cause diminished river flows with consequent damage to salmonid habitat. It is claimed that diminished river flows cause increased siltation, low survival of 0-group trout and salmon, and downstream displacement of spawning activity (Giles *et al.*, 1991); a number of English rivers suffering from, or threatened by, overabstraction has been listed (Environment, Agency, 1998a). One example is the River Piddle (Dorset, UK). This is a chalk river fed chiefly by chalk springs and is important as habitat for resident and migratory salmonids. Investigations showed that abstractions from boreholes caused an unacceptable loss of flow in the upper parts of the river. The PHABSIM model (*see* Chapter 6) was used to show that the abstraction would lead to loss of juvenile habitat during the period May to October inclusive and the views of a panel of expert anglers suggested that the abstraction reduced the occurrence of 'good' fishing days by about 21 days per year (Strevens, 1999). Steps have now been taken to reduce abstraction during periods of low flow (Environment Agency, 1997a). A general review of the system of water abstraction licencing in England and Wales is in progress (Department of the Environment, Transport and the Regions, 1998).

5.2   Fishing and fishery management

Man's most direct effects upon salmonid populations are through cropping by legitimate means, such as licenced rod fishing for sport or licenced commercial fishing by net or fixed engine, and through a variety of illegitimate poaching methods (some of them very destructive). Humans, in their capacity as anglers, fishermen or poachers, can be regarded as yet another predator whose behaviour can give information that is useful in framing rules for the regulation of legitimate fishing and the control of poaching. As an example, we will consider angling for salmonids, with special reference to trout in fresh water. In Great Britain there are three main species that are angled for sport: the two native species of *Salmo* and the rainbow trout *(O. mykiss)*. The two species of trout are angled in both still and flowing water. In flowing water the fishing is mainly for brown trout in fisheries composed either entirely of indigenous fish or of varying mixtures of wild and stocked fish. In general, the importance of stocked fish is greatest in the south and east, and decreases to the north and west. Some still waters, especially in the north and west, contain only wild brown trout but the populations of others are increasingly being augmented by hatchery fish, especially rainbow trout. A proportion of still-water fisheries is run entirely on a 'put-and-take' basis, with an increasing emphasis, recently, on stocking at heavy rates with rainbow trout.

During the last two decades there have been a number of studies aimed at gaining better understanding of the quantitative relationships between humans and trout in these various types of fishery. In the short term the catch rate in put-and-take fisheries (and probably in others) is influenced by weather conditions (Taylor, 1978). A number of studies (Fleming-Jones & Stent, 1975; Crisp & Mann, 1977a; Pawson, 1982; North, 1983) have shown that given a reasonably consistent temporal pattern of stocking with fish of a relatively standard size, there are simple positive relationships between the numbers of fish stocked each season and the catch by anglers. There also appear to be negative relationships between the numbers of fish stocked and their growth (if any) after stocking. In natural fisheries there is evidence of similar relationships. One study (Engstrom-Heg, 1986) has shown that the average catch per hour of wild brown trout in a New York (USA) stream can be predicted from the population density of yearling trout in the previous autumn and the fishing intensity (hours fished per unit area of stream). Similarly, in Cow Green Reservoir (UK), the angling catch of brown trout (chiefly age group III) in any year was correlated with an index of the abundance of trout of ≥ 15 cm length (mostly age group II) in the previous year, which in turn was correlated with an index of abundance of trout of < 15 cm length (mostly age group I) in the previous year (Crisp *et al.*, 1990). In both of these examples, one for flowing water and the other for still water, knowledge of the abundance of young fish could be used to predict angling catches one or more years ahead.

Analyses of the frequency distribution between anglers of the numbers of fish caught per angling trip show clearly that in most fisheries a large proportion of the catch is taken by a small percentage of the anglers. For example, at Cow Green Reservoir in 1972, 51% of the catch was taken by 9% of the anglers and most angler-visits accounted for no catch (Crisp & Mann, 1977b). Such analyses can be used to assess the effect, on saving fish, of imposing a bag limit of any specified number of fish in both flowing water (Alabaster, 1986) and still water

(Crisp & Robson, 1982). Such studies on the impacts of humans as predators can be of value in the management of fisheries. They are probably capable of much further development in the future. Their basis depends, however, on the collection of sound catch statistics by the managers or owners of fisheries and upon accurate reporting by anglers. Otherwise the scope for useful analysis is limited.

There has been concern in recent years not only about a general decline in salmon runs in some rivers, but also about a decline in the numbers of 'large spring salmon'. Some of the difficulties in getting to grips with the problem arise from vagueness of definition. The most simplistic view, adopted by many anglers, is that these fish are large salmon that run only in the early part of the season, that they form a relatively separate breeding sub-population within each river system, and that they tend to breed true. This view leads to the suggestion of relatively simplistic solutions aimed at conserving more large spring salmon to spawn and/ or enhancing their numbers by artificial propagation. It would be more accurate to describe these fish as 'early-running, multi-sea-winter salmon' and to note that some large multi-sea-winter salmon run later in the season whilst some grilse may run early in the season and may interbreed with the large early-running fish. We need, therefore, to consider two separate traits. The first is the tendency to run early in the season and the second is the tendency to remain at sea for more than one winter. These two do not necessarily have the same causation, and are not necessarily correlated. A number of possible causes for the decline has been put forward including:

(1)   that this trend is part of a natural cycle and will eventually be reversed;
(2)   that it reflects environmental changes in the sea;
(3)   that there is over-exploitation of large spring salmon;
(4)   that it can be related to the disease known as 'Ulcerative Dermal Necrosis' that killed many salmon during the 1960s;
(5)   that it arises from habitat degradation in the rivers.

There is evidence of some genetic influence upon the seasonal return pattern of Atlantic salmon (Hansen & Jonsson, 1991) and there is, thus, possible merit in seeking to conserve or enhance the sub-population of early-running fish, even though there is little reason to believe that they do not interbreed with later-running fish. There is also a tendency for fish running early to move greater distances upstream than those returning later (Salmon Advisory Committee, 1993), and possibly for them to spawn selectively in certain tributaries. We know that in the coho salmon (*O. kisutch*) faster freshwater growth gives larger smolts, which have a higher probability of returning as one-sea-winter fish (Bilton *et al.*, 1982; Mundie *et al.*, 1990). There is also evidence of a negative correlation between smolt size and sea age at maturity in the Atlantic salmon (Randall *et al.*, 1986; Nicieza & Brana, 1993). Two important possibilities arise from this. First, the production of large smolts in hatcheries as part of 'stock enhancement' programmes may be counter-productive insofar as it may lead to an increased proportion of grilse amongst the returning adults. Second, any process of nursery stream degradation that acts selectively against those tributaries tending to produce old smolts will also tend to reduce the proportion of adults that return as multi-sea-winter fish. This is a complex problem and we still do not fully understand the reasons for the decline. A great deal more research is needed especially into the relative contributions of heredity and environment to the production of early-running and multi-sea-winter fish.

Inspection of stock-recruitment curves (Figs 2.6 and 2.7) shows that for populations whose numbers are usually regulated by density-dependent mortality (represented by the line b–c in Fig. 2.7), man can often harvest a substantial fraction of returning adults and this will lead to a decline in sub-sequent generations only if the numbers harvested reduce the parent cohorts to levels that correspond to the sloping part of the curve (a–b in Fig. 2.7). The numbers of adults allowed to spawn should be sufficient to ensure full utilization of the available spawning and juvenile habitat. In practice, there is considerable scatter of data points around the theoretical curves and this reflects appreciable between-years variation in mortality from density-independent causes, such as droughts. On the 'precautionary principle', some allowance must be made for unpredictable variations of this type. Ideally, application of this approach would require sound scientific data on salmonid stocks and population dynamics in individual rivers (or even individual tributaries). For each river this would consist of estimates of spawning population (from electrofishing surveys of resident fish and from the use of counters/traps/tagging for migratory fish) and a stock-recruitment curve for each relevant species. In practice, we have fish counters on relatively few rivers in the UK and most of the published stock-recruitment curves are for rather small streams. Sound and relatively comprehensive data exist for a few larger systems (e.g. the River Bush in Northern Ireland: Kennedy & Crozier, 1995; Crozier & Kennedy, 1995) but it is unlikely that application to other rivers of the stock-recruitment curves generated by these studies can be justified, except as a very approximate guide. Considerably more work is required in the collection and analysis of long-term data sets from a variety of river types before it will be possible to set meaningful 'spawning targets' for individual rivers. Meantime, the fishery manager will be required to make rela-tively arbitrary decisions on appropriate catch levels.

Once a decision has been made as to what catch can be allowed, while still conserving the population on a long-term sustainable basis, there is a need to share the catch between different groups of participants (e.g. rods and nets) within the fishery. Sharing of the catch is usually achieved by the use of licences and regulations that are aimed at regulating fishing effort and, hence, catch. These measures involve social, political and economic factors and therefore contain a substantial element of subjective judgement. Substantial consultation and compromise may be needed to ensure that the enforcement of fishing regulations is seen to be fair and yet is effective and not unduly expensive.

Further potential threats to the well-being of wild salmonid stocks arise from the increasing development of intensive rearing units both in fresh water for brown and rainbow trout and in the sea for Atlantic salmon. These ventures can be very profitable and they generate valuable employment. The industry appears to have grown and spread rapidly in recent years. It is likely that the legislative framework, the regulatory mechanisms and our knowledge of the broader ecological consequences of these units have lagged behind this development. Freshwater hatcheries and fish farms usually abstract water from springs or streams and then release effluent to the same stream. The abstraction point is generally upstream of the release point. If the abstraction is large relative to the total stream discharge, then the portion of stream between the points of abstraction and release may have greatly reduced flow and this may impede the spawning movements of wild fish. The effluent from such a unit may differ in temperature and pH from those in the natural stream and it may be contaminated with materials used for disinfection and disease control within the unit. It is also likely to contain the metabolic products of the fish and, probably, decomposing waste fish food. These materials are likely to give an increased oxygen demand, increased suspended solids and enrichment of part of the recipient stream. Ideally, there should be insistence that when such facilities are set up, the outflow shall be upstream of the inflow so that the units are, in effect, fed by their own effluent, thus giving a powerful incentive to treat the effluent adequately and also avoiding problems of reduced flow in part of the stream.

By their very nature, fish farms and hatcheries lead to frequent transportation of fish in and out. This, and the intensive nature of the cultivation, give a high risk that diseases and/or parasites will eventually occur and proliferate, spread rapidly and may reach the wild salmonids outwith the unit. In addition, short of the installation of some sort of expensive and dangerous electrical barrier that is lethal to fish, it is difficult to prevent some escapes, even from a very well run freshwater hatchery and; losses from marine facilities can often be surprisingly large. During the autumn of 1998, an estimated 10 000 salmon escaped from sea cages in or near the Firth of Clyde (Scotland) when a cage was damaged while being moved and in November 1998, an estimated 17 000 escaped from a fish farm near Oban (Scotland). The effects of such escapes from marine facilities will be considered below.

In Norway, parts of Scotland and Ireland and elsewhere, salmon are now reared intensively in cages in fjords and sea lochs, and sometimes in the open sea. In a similar manner to freshwater rearing facilities, this cage culture might be expected to cause problems from the decomposition of uneaten food on the sea bed and from the application of medication and prophylactics and their consequent effects on the marine biota. As in most intensive rearing processes, there are problems with diseases and parasites. The sea louse (*Lepeophtheirus salmonis* Kröyer) appears to be an important pest, and it has been claimed that some of the chemical methods used to counter this parasite may harm nearby populations of crabs and lobsters. In the late 1980s, there were indications of a sharp decline in sea trout stocks in some areas. The fish were returning from the sea earlier and in smaller numbers than usual. They were emaciated and carrying unusually large burdens of sea lice. The main areas concerned were parts of the west coasts of Scotland and Ireland and there was a substantial (but not complete) spatial coincidence of the areas of most intensive salmon farming and the areas of marked sea trout decline (Department of the Marine, 1993). This sort of coincidence does not prove a causal link, especially as some aspects of the recording and analysis of the Irish data have been criticized. It does, however, give cause for concern. Pending clarification of the complex issues raised by salmon farming, it is arguable on the precautionary principle that a moratorium should be placed on any further development of the industry until we have sufficiently detailed knowledge to take measures to mitigate or prevent any further damage.

A virus disease known as infectious salmon anaemia was observed in salmon farms in Norway in the mid 1980s and in Canada in the mid 1990s. This disease caused substantial losses of farmed fish. In 1998, it was observed in some Scottish units and, in association with a collapse of prices as a result of overproduction, it has had a serious impact on the industry. Under European Union regulations all stock must be slaughtered at a farm when an outbreak is confirmed. The manner in which this disease reached Scotland is not yet clear.

The production of brown trout, rainbow trout and Atlantic salmon in intensive units leads, via escapes or via direct stocking, to some of these fish entering natural watercourses. Deliberate stocking with hatchery-reared salmonids has a long history and can be valuable in reintroducing the species to places where, for one reason or another, they have been extinguished. The rearing of salmon for release at the eyed egg or parr stage is a method often used in attempts to enhance salmon stocks. The benefits of much of this stocking are dubious. It is likely to be wasteful and counter-productive in waters that are already well-stocked, and where it does contribute to natural production in the form of extra returning adults, the financial cost per returning fish may be very high. The release of hatchery-reared smolts makes a major contribution to some commercial fisheries for *S. salar* (Thorpe, 1980).

Programmes of stocking with reared fish, accidental escapes from rearing units and the transfer of fish between regions are all accompanied by genetic implications. The salmonids of individual river systems (or even individual tributaries) are adapted genetically to the conditions in that river/tributary. The escape or deliberate release of hatchery/farmed fish carries a risk of diluting the genetic make-up of the indigenous population (Cross, 1989; Reisenbichler, 1997; Crozier, 1998) so that they are less well adapted to their environment. Such impacts of the release and/or escape of farmed Atlantic salmon on indigenous populations are not yet fully understood but are leading to increasing concern, especially in Norway (Hvidsten *et al.*, 1994; Einum & Fleming, 1997; Fernö & Järvi, 1998). A particularly interesting example of these genetic problems is that of *Gyrodactylus salaris* Malmberg. This is a skin parasite that originated from the Baltic drainages of Sweden where it appears that host and parasite are in balance, presumably because they have, over a long period, became genetically adapted to one another. This organism appeared in Norwegian salmon farms in the 1980s and subsequently spread to rivers (Mills, 1989). The parasite can cause severe reductions in the numbers of young salmon in river systems whose salmon populations are not genetically adapted to it (Johnsen & Jensen, 1986) and it can, apparently, be controlled only by complete elimination of young salmonids from the river followed by restocking. By 1997, it was widespread in Europe and had been recorded from Sweden (where it is native), Norway, Russia, Denmark, Germany, France, Spain and Portugal. The possibility of its transfer to the UK is being taken seriously by the relevant authorities.

The next major development may be the process of 'ocean ranching' in which hatchery salmon smolts are released, unprotected, into marine waters and are then harvested when of marketable size. Thorpe (1980) examined the issues and concluded that current salmon farming methods were biologically wasteful because they were nett consumers of high-grade protein, whereas ranching would be biologically valid in this respect. He commented, however, on the need for development of sound management techniques and noted the need for a new legal foundation to protect the ranching enterprises. Perhaps we should also consider the need for a new legal foundation to protect the indigenous salmonid populations from the activities of the prospective ranching interests?

The easy propagation and fast growth of the rainbow trout make it popular with the managers of hatcheries and stocked fisheries. It is also an adept escaper and, as a result of direct stocking and escapes from rearing units and stocked fisheries, it is now widespread in Great Britain. In 1940, the rainbow trout was recorded in between 50 and 56 waters in Britain (Worthington, 1940, 1941). Thirty-one years later the species was recorded in 462 localities (Frost, 1974) and was said to breed naturally at about 40 of them (though at only three of them was reproduction considered to give a self-perpetuating population). The three self-perpetuating populations were all in streams with high pHs and equable temperature regimes. During the following two decades, the distribution of rainbows has widened substantially and their numbers in our lakes and rivers have also increased. Jowett (1990) concluded that rainbow trout were favoured by spring-fed rivers with equable flow and temperature regimes. On this basis, suitable habitat for rainbow trout to establish themselves readily will exist in only a limited number of British rivers.

The rainbow trout is proving to be a very adaptable species. It is particularly plastic in its spawning time and appears to grow more quickly in fresh water than does either species of *Salmo*. There is at least one population that is probably self-perpetuating in a flashy stream with comparatively low winter temperatures in northeastern England and there is evidence from sightings of smolts and from captures of larger fish at sea that a degree of anadromy has been regained. Sægrov *et al.* (1996) recorded natural reproduction of anadromous rainbow trout in Norway and considered that this might represent a threat to indigenous species. Kocik & Taylor (1994) studied streams feeding the Great Lakes in North America and concluded that there was little evidence of *O. mykiss* causing a decline of *S. trutta* populations, but they did advise fishery managers to exercise care when seeking to manage the two species together. In the alpine Rhine valley, introduced rainbow trout have begun to reproduce naturally at some time during the last two decades and they have invaded tributaries and lakes. Analyses of the present status of salmonids (Peter, 1995) show substantial replacement of brown trout by rainbow trout to the point where, in some streams, rainbow trout contribute over 90% of total trout numbers and the brown trout is considered to be endangered. We should, therefore, regard the release of *O. mykiss* warily. Future release of other exotic species should be strongly discouraged, at least until the species in question is proved to be benign beyond any reasonable doubt.

5.3   Industry

The effects of industrial activity upon the aquatic environment are many and varied. It is possible to give only brief mention to a representative selection. Two distinctions are useful. The first is between past and present industries. Many present-day industries have, from their inception, been obliged to meet stringent restrictions on the quality and quantity of their effluents and upon their working practices, while those older industries still in operation are, generally, becoming more environmentally conscious. Unfortunately we still have a substantial legacy from past industrial activity in the form of continuing pollution from derelict land and waste tips and various obstructions to fish movement. The second distinction that can usefully be made is between extractive and manufacturing industries.

The main extractive industries include quarrying, sand and gravel extraction, coal mining, china clay production, and mining for various other minerals. Some of these activities can cause substantial increases in the suspended solids of streams and, even when the solids are inert, they can still be harmful to salmonids (Alabaster, 1972). In addition, the products of some mineral extraction processes can be toxic. The extraction of gravel from rivers and streams during the spawning and incubation periods of salmonids may not only damage spawning sites but may also cause suspended solids problems downstream. Past mining for heavy metals in many parts of upland Britain has left a large amount of mine spoil. Reworking of this spoil either by heavy rain and large spates or by human attempts to recycle it may release high concentrations of metals such as zinc and lead. In recent years a large number of deep coal mines have been closed in

Britain. With hindsight, the process would appear to have been carried out with insufficient consideration of the environmental consequences. For example, there appears to have been inadequate provision to ensure that a body was nominated, empowered and properly funded to continue the pumping of water from these mines after they ceased to operate. The water pumped from working coal mines is generally of acceptable quality for discharge to watercourses, but when pumping ceases the water often becomes acidic and rises slowly through the abandoned galleries and shafts and takes into solution iron and a variety of other materials before finally reaching the river systems as a highly toxic solution.

Past manufacturing industry has left a legacy of weirs and similar river obstructions and also some polluted land that has the potential to produce leachates that may harm the aquatic environment. Some modern industries produce rather intractable effluents. Examples include metal industries (ammonia, phenols, metal, oil, heat), chemical industries (many and varied pollutants, including heat), electricity generation (chiefly heat and strong acids from the burning of fossil fuels but others, including nuclear pollution, are possible), textiles (dyes and pesticides), food manufacturers and distilleries (heat and biochemical oxygen demand) and timber and paper industries (suspended solids and possibly some chemical pollutants). It is worth noting that even well-managed industrial activities can lead to environmental problems as a result of accidents. Such accidents often arise from continued use of outdated and worn out pipework and other equipment, or from the actions of inadequately trained or supervised individuals, or simply from a lack of adequate care.

5.4   Transport and roads

Simple unsurfaced farm and forest roads, particularly during and soon after construction, can be a source of suspended solids in streams. If not constructed and used carefully they can lead to continuing problems of land erosion in some upland areas. Similar problems can arise from thoughtless use of farm plant, and from the use of off-road vehicles for sport and, in heavily used upland areas, from walkers and mountain bikers.

The effects of urban roads and motorways during their construction and use have been considered in detail by the Royal Commission on Environmental Pollution (1992a) and most of the quantitative information given below is from that source. During construction of one motorway, suspended solids concentrations of 22–$336\,\mathrm{mg\,l^{-1}}$ (mean $86\,\mathrm{mg\,l^{-1}}$) were recorded in one recipient river. In addition to soil, the construction of surfaced roads can lead to harmful releases of cement liquors and phenolic compounds from road surfacing materials. The rate of run-off from road surfaces is usually more rapid than that from vegetated ground and suspended solids concentrations in road run-off in the UK can reach 1400–$2000\,\mathrm{mg\,l^{-1}}$ on urban roads, and $5500\,\mathrm{mg\,l^{-1}}$ on motorways. Most of these solids are soil particles which include up to 25% organic material and this can lead to biochemical oxygen demands of 100–350 mg l^{-1} in urban run-off and a mean value of 32 mg l^{-1} in motorway run-off. Road run-off also contains variable amounts of lead, zinc and other metals. Contamination and/or enrichment can arise from de-icing materials. On roads the main de-icer is salt, though urea is occasionally used, and rates of application of 35 g m^{-2} have been quoted. Airport de-icing usually uses glycol or urea and both of these create an oxygen demand in the run-off water.

Much of this material from the roads finds its way, untreated, into watercourses. The impact from some surfaced rural roads is likely to be very substantial and will be mentioned later in this chapter.

In addition to these relatively routine sources of pollution, the road system may generate a variety of problems arising from road accidents and consequent spillages of fuel, battery acid and miscellaneous freight. A surprising number of major roads run alongside lakes and rivers. The herbicides used beside roads and railways may be a further cause of aquatic pollution.

5.5   Urban development

Some aspects of urban development have already been considered in this chapter. The development of urban areas increases the rate of run-off of rainfall and increases the risk of pollution of that run-off as it moves towards the river system. Perhaps the aspects of gathering together large numbers of people in urban areas that are most relevant to salmonids are water consumption, production of sewage, production of other waste, atmospheric pollution, and miscellaneous accidental pollution episodes.

The presence of large numbers of people inevitably means that there will be a high demand for water, and in areas where population density is high and precipitation is low, this can lead to problems of over-abstraction. Some domestic uses of water, such as bathing and drinking, are relatively benign because most of this water returns to the river system via the sewage works. Costs are incurred in treatment and the water does not return to the same part of the river system (or necessarily to the same river) as that from which it was originally taken. Most of it does, nevertheless, return to the aquatic system. In contrast, much of the water used for such purposes as the watering of gardens, lawns, bowling greens and golf courses will go straight into the atmosphere, through evapo-transpiration, and be lost to the river system.

Both urban and rural communities produce sewage. In Britain this is treated either in septic tanks serving one or several dwellings, or most often by sewage treatment plants serving whole communities. Both systems can be effective if constructed and used properly. A problem arises from the existence of 'combined' systems of sewerage in which storm water from streets passes through the same pipes as the domestic sewage. In periods of heavy rainfall such systems can become overloaded and may discharge untreated sewage into river systems. A further problem comes from those systems that do not fully treat the sewage and release an effluent with a high oxygen demand.

The disposal of domestic and industrial waste is a growing problem. Much of it has been placed in landfill sites and this practice, particularly when poorly planned and supervised, can store up troubles for the future as various noxious leachates find their way into rivers.

In urban areas large amounts of fossil fuel are consumed in industry, domestic heating systems, and vehicles. The main product of this combustion is carbon dioxide, but sulphur dioxide and various oxides of nitrogen are also produced and the two latter are the main acidic components of 'acid rain'.

England has a projected need for 4.4 million new homes between 1991 and 2016. This new housing will be a major contributor to a projected 30% increase in water demand and will put pressure on water supply and sewage treatment plants.

5.6   Forestry

Woodland covers approximately 10% of the land area of Britain (Fig. 5.1). Approximately 70% of this is conifers (Forestry Industry Council of Great Britain, 1996) and the remainder is broadleafed trees. Much of the coniferous woodland is man-made forest and much of that has been planted in upland areas whose streams are important spawning and nursery areas for trout and salmon. Most of this section is focused on the effects of this conifer production.

During the planting phase it is usual to construct roadways and, often, to cut furrows for tree planting (Fig. 5.10) and to cut drains. It is also usually necessary to apply fertilizers and, possibly, herbicides when the trees are planted. The possibility of pesticides entering watercourses is obvious. The cutting of drains may, however, have much greater and longer-lasting effects. The first is to increase the speed of run-off. The second effect of drain-cutting and of the construction of roads and culverts is to increase the rate of soil erosion, and hence the concentrations of suspended solids in the watercourses. Drains that are cut through peat into the underlying mineral soil may lead to severe and continuing erosion of that mineral soil (Fig. 5.11).

As the trees grow they form a 'closed' (continuous) canopy and this causes shading, loss of water yield, some acidification and sometimes a reduction in soil erosion. Shading may reduce the available light, and hence reduce the growth of stream algae. It will also influence stream temperatures. Relative to grassland or open moorland, coniferous forest loses a larger percentage of input precipitation by transpiration and evaporation from the canopy and this gives lower stream discharges in forested streams than in similar unafforested streams. Coniferous

Fig. 5.10 Forest ploughing up and down the slope. This is likely to increase the rate of run-off and the rate of erosion. Mid-Wales, c. 1984. Photograph by Prof. Malcolm Newson.

trees, on their own account, have some capacity to acidify the soil in which they grow. They do this by absorbing positively charged ions, such as calcium, through the roots and releasing hydrogen ions from their leaves. These acids may then be washed down the outsides of the trees during rainfall (Table 5.4). Where this is the only mechanism at work, and particularly where the soil contains sufficient calcium and other bases to neutralise the effects of this acidic downwash, there is no real threat to the aquatic environment. Serious problems do arise, however, when additional factors come into play. The burning of fossil fuels in power stations, homes and vehicles generates sulphur dioxide and oxides of nitrogen that dissolve in water to give strong acids. These strong acids can be deposited on the needles and stems of conifers as 'dry deposition' (to be washed down later by rain) and may also fall directly in the rainfall. If the underlying soil is low in buffering materials, such as calcium, the inputs of acidic water during heavy rainfall may pass rapidly through or over the soil and appear in the recipient stream as 'acid pulses' or 'acid episodes'. Problems from so-called 'acid rain' are most apparent in those northern and western parts of Britain that have hard, non-calcareous rocks. These areas carry much of the man-made forest and are very important areas for trout and salmon. Aluminium is a widespread and abundant element in nature and can appear in several different ionic forms ('species'). It forms a major component of clay and, under most circumstances, is harmless to fish. In acidic (low pH) water some of it is converted to the form Al^{+++} which at very low concentrations can kill salmonids (Table 3.4). During this lengthy phase of forest growth, various maintenance activities, such as thinning and further

Fig. 5.11 A forest drain about 50 years old. The drain was originally 30–50 cm deep, but was cut into the underlying mineral soil and this led to severe erosion. Wales 1998. Photograph by Phil and Sue Hill.

Table 5.4 pH values, weighted for flow, for rainwater passing through a coniferous tree canopy. 'Rainfall' refers to the input rainwater, 'throughfall' refers to rainwater falling to the ground after passing through the tree canopy, and 'stemflow' refers to water flowing down the tree trunk. Derived from Neal *et al.* (1992).

	Minimum	Mean	Maximum
Rainfall	3.4	4.8	7.7
Throughfall	3.6	4.5	6.1
Stemflow	3.2	3.9	4.6

applications of fertilizer and pesticides, may be needed. The thinning may require the deployment of heavy equipment and this may cause soil erosion.

The final stage of forestry is that of felling, removing the timber, clearing-up the site and, probably, replanting. This work may lead to further episodes of high suspended solids and production of chemical leachates. This final stage of the process has been little studied in the UK.

Most of the detailed studies of the effects of forestry activity upon salmonid fishes have been made in North America and not all of the findings are capable of being imported, uncritically, into Britain. This is a consequence of differences in both geography and scale. Fortunately, however, recent years have seen the publication of summaries of detailed long-term hydrological studies on effects of forestry activity in Wales and other parts of the UK (Kirby *et al.*, 1991; Neal *et al.*, 1992; Robinson *et al.*, 1994) and the publication of results from shorter term studies on forestry and acidification elsewhere in Wales (Stoner *et al.*, 1984; Ormerod *et al.*, 1989; Weatherley & Ormerod, 1990). In the light of this and other information, it is possible to quantify some of the effects associated with forestry activity and to relate them to the needs of trout and salmon. It is important to note that some of the quantitative information may not be universally applicable and that the correlations that are apparent do not necessarily imply causality. In fact, we still have only a very limited understanding of the details of many of the mechanisms that are at work.

Pre-afforestation drainage work in a small catchment in northern England led to a much sharper hydrograph peak and to the peak being attained two to three hours earlier than before drainage and this effect was still apparent six years after planting (Robinson, 1980). As the trees grow and the forest matures, the effects of initial drainage works decrease but the amount of water loss through evapo-transpiration increases. Between 1975 and 1985 in the upper Wye, which is sheep pasture, about 15% of the rainfall was lost by evapo-transpiration. The loss in the afforested upper Severn was 23% (Kirby *et al.*, 1991). Suspended solids concentrations in run-off during the draining and planting phase can be high. In a small catchment in northern England, concentrations of suspended solids averaged $4\,\mathrm{mg\,l}^{-1}$ before draining. During ploughing, values ranged from

30 mg l^{-1} in dry weather to 150 mg l^{-1} in rainy weather. After hand-clearing of blocked drains, values generally ranged from 300 mg l^{-1} to 1700 mg l^{-1} but with maxima up to 7000 mg l^{-1} (Robinson *et al.*, 1994). These very large quantities are likely to be harmful to salmonids (Table 3.4). A comparison between two small catchments that are similar in all respects apart from land use (Table 5.5) shows a markedly higher mean suspended solids content in an afforested stream than in one draining pasture. The mean suspended solids concentration in the former more than doubled during the felling process. Comparison with Table 3.4 shows that the mean concentrations were within the 'preferable' range in the streams draining pasture and mature forest, but they fell into the 'acceptable' range during forest felling. Concentrations are, of course, likely to have been well above the mean values during individual spate events, especially during forest felling. There is also evidence that the movement of 'bed load' (gravel framework material moving along the stream bed during large spates) is greater in afforested than in moorland streams. Mean annual values of 11.8 and 38.4 tonnes km^{-2} year^{-1} were estimated for two streams draining afforested catchments and a value of 6.4 tonnes km^{-2} year^{-1} for a stream draining a moorland catchment (Kirby *et al.*, 1991). This implies a greater risk of washout of intragravel stages of salmonids in the afforested than in the unafforested streams.

Table 5.5 Mean concentrations of suspended solids in a stream draining sheep pasture and an afforested stream before (a) and during (b) clear felling. Derived from Kirby *et al.* (1991).

Catchment	Area (ha)	Land use	Suspended load (mg l^{-1})
Afon Cyff	3.13	Pasture	3.0
Afon Hore (a)	3.08	Mature forest	13.0
Afon Hore (b)	3.08	Mature forest being felled	30.5

It is generally agreed that the main effect of afforestation upon stream water temperatures is to damp down the daily and annual fluctuations. The felling of deciduous woodland around one small stream led to increases of summer water temperature by up to 6.5°C (Gray & Edington, 1969). The main effects of coniferous forest are to decrease the daily fluctuation, to lower summer temperatures (chiefly by lowering daily maxima) and, in some instances, to raise winter temperatures. There are detailed analyses of data from two studies in Britain (Weatherley & Ormerod, 1990; Crisp, 1998); both studies showed a similar depression of summer temperatures in afforested streams though the former suggested a small elevation of winter temperatures and the latter did not. In both studies the summer depression of monthly means in the afforested streams was less than 2°C. Weatherley & Ormerod found that the observed effects of afforestation modified the dates of swim-up of *S. trutta* by a few days and could substantially reduce growth rate during the first two seasons of life.

Crisp & Beaumont (1998) found that the effects of afforestation on water temperature had negligible effect on predicted swim-up date (about three days delay) but reduced the predicted mean instantaneous growth rate per day over a year for a trout starting at 6.0 g weight on 1 January by 5% in 1993 and by 10% in 1994. The evidence available to date suggests that in British streams, the effects of coniferous forest on water temperatures are relatively slight. The changes may reduce salmonid growth rates but do not appear likely to have a direct influence on salmonid survival.

Ormerod *et al.* (1989) made a wide-ranging study of aluminium concentrations and species in a large number of afforested and unafforested streams in Wales. They found pronounced correlations between percentage of coniferous forest cover, pH (negative correlation) and aluminium concentration (positive correlation), and found that pH was a useful predictor of aluminium concentration. They did, however, make the reservation that such correlations, although suggestive, do not necessarily imply causation. In a less wide-ranging study they found that when the concentration of aluminium was increased under forest, most of it was present in the labile form (Al^{+++}). Recent analyses of the fluctuations of pH and aluminium concentration in the Institute of Hydrology's experimental catchments at Plynlimon (Wales) give a good illustration of the effects of afforestation upon aluminium concentrations and species. The catchments of the upper Wye and upper Severn are very similar in all respects except that the former is sheep pasture and the latter is largely covered by mature forest. Comparison between two sub-catchments of the upper Wye (Afon Cyff and Afon Gwy) and two sub-catchments of the upper Severn (Afon Hore and Afon Hafren) show similar values of pH (6.0–7.0) and aluminium concentration ($< 0.05\,\mathrm{mg\,l^{-1}}$) in all four streams during base flow (Fig. 5.12). During storm flow, however, pH was lower and dissolved aluminium concentration higher in the afforested than in the moorland streams. It is also apparent (Fig. 5.13) that the storm flow pH values, especially in the afforested streams, enter the range in which an increasing proportion of the aluminium present takes the Al^{+++} form that is considered likely to be harmful to salmonids at quite low concentrations (Table 3.4). Neal *et al.* (1992) concluded that the concentrations of environmentally harmful forms of aluminium were likely to increase further following deforestation and this is supported by comparisons within the Afon Hore, before and after clear felling.

5.7   Agriculture

Agriculture in various forms occupies over 75% of the land surface of Britain and its impacts upon salmonid fisheries are of major importance. It covers a wide range of activities many of which impinge directly or indirectly upon the environment of trout and salmon. The significance of its various effects may vary not

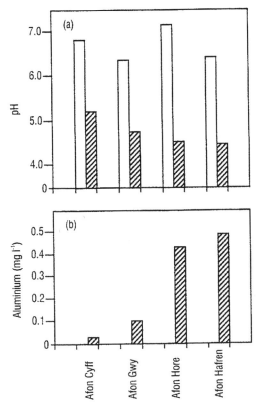

Fig. 5.12 (a) Typical values of pH in base flow (plain bars) and storm flow (shaded bars) for two moorland streams (Afon Cyff and Afon Gwy) and for two afforested streams (Afon Hore and Afon Hafren). In all four streams aluminium concentration during base flow is $<0.05 \, \text{mg} \, \text{l}^{-1}$; values during storm flow are shown as shaded bars in (b). Derived from Crisp & Beaumont (1996) but based on an unpublished report.

only between different types of agriculture but also between different regions, particularly as a result of differences in climate and soil type. It is therefore convenient to consider the possible impacts of agriculture as six general topics and to give some comments on, and/or illustrations of, each. The six topics are listed in Table 5.6, together with a brief indication of the ways in which each one may become a problem.

Some land drainage is necessary for effective land use in both uplands and lowlands. In the past, drainage schemes in headwater areas were encouraged with subsidy and in some areas of England and Wales they are still fostered by 'Internal Drainage Boards'. Such schemes generally give rapid run-off of winter precipitation but may also cause flash spates, high suspended solids concentrations, and reduced aquifer recharge and low flows during summer. As a general guide it is likely that over-enthusiastic and unsympathetic land drainage (including moor draining) will be harmful to salmonids. The situation is, how-

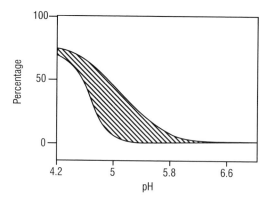

Fig. 5.13 Relationship between pH and the percentage of aluminium present in the form Al^{+++} in the Afon Hafren. These are field results and all of the observed values fell within the shaded area. After Neal *et al.* (1992).

Table 5.6 Six general topics relating to the influence of agricultural activities upon the aquatic environment and, hence, upon salmonid habitat.

General topic	Details
Flow modification	Land drainage especially by surface ditches gives more rapid run-off, reduced summer flows, and higher suspended solids. Irrigation of land causes reduction of flow in donor river system.
Poisons	Pesticides (including sheep dips) and herbicides may enter watercourses during or after storage, use, abuse, accident or disposal.
Inorganic enrichment	Fertilizers can be damaging through overuse, use too close to streams, and entry of streams by windblow or washout.
Organic enrichment	Slurry, silage and manure can cause problems through leakages from storage, wash-in from roads and fields, and application of slurry too close to streams or just before heavy rain (especially when the ground is frozen).
Suspended solids	Can enter streams from water and wind erosion of land (aggravated by ploughing unsuitable land or by overstocking with livestock), deposition of mud on roads, passage of agricultural vehicles through water courses, and bank damage caused by agricultural traffic and animals.
Structural damage	Stream 'training' (e.g. dredging and straightening) and gravel extraction cause increased suspended solids and probably instability of the stream bed and channel and loss of salmonid habitat. Vehicle crossings cause increased suspended solids, possible compaction of bed and mechanical shock to intragravel stages of salmonids.

ever, relatively complex. A detailed review of evidence (Robinson, 1990) showed that the precise effects of drainage depend both upon the type of drains and upon soil type. At the field scale, open ditches generally give higher spate peaks than do sub-surface drains. Pipe drains may even decrease the rate of run-off. At a catchment scale, ditching and channel 'improvements' usually lead to more rapid run-off and higher channel velocities.

It is arguable that some drainage schemes have, by reducing the recharge of aquifers and the summer levels of rivers, aggravated the conflicting demands for water during periods of drought. It is at such times that farmers have the greatest need to irrigate crops and a common method is by 'spray irrigation' in which water is ejected into the air as a jet or spray. This method uses large amounts of water at times when surface waters and aquifers are most stressed and it is wasteful in the sense that much of the water returns to the atmosphere via evapo-transpiration. 'Trickle irrigation', in which the water is supplied directly to the plants via pipes or other apparatus, is a less wasteful alternative.

Herbicides and pesticides are widely stored and used in agriculture, often in large quantities. They can reach watercourses by windblow, leaching, with eroded soil and through miscellaneous mischances arising from careless storage or use, accidents and thoughtless/irresponsible disposal. Some of the solvents and carriers in which these materials are applied can be just as harmful as the 'active ingredients'. Sheep dip – particularly its careless disposal – has always presented a serious threat to streams and rivers. In the past the main basis of such dips was organophosphates such as lindane and dieldrin. These were persistent and very toxic to vertebrates including humans, sheep and fish. As a result of concern about their effects on human health, organophosphates have now been largely replaced by synthetic pyrethroids. These are less harmful to vertebrates but much more harmful to invertebrates, including most of the prey items of salmonids. The Environment Agency stated that the figures for sheep dip incidents in England and Wales in 1997 were the worst on record, causing widespread damage to hundreds of kilometres of rivers, especially in northwestern England, the English Midlands, and Wales. Efforts are currently being made to tighten up controls on the disposal of sheep dip. Disposal of these materials in soakaways or on 'sacrificial land' (from which they are likely eventually to reach watercourses) is not satisfactory. Removal of all used sheep dip for responsible recycling or disposal is an attractive but probably impractical measure. Other possibilities include increased use of mobile dipping facilities so that the responsibility does not rest with individual farmers.

In recent decades the British yields (tonnes ha^{-1}) of most crops have risen steadily as have the annual inputs (kg ha^{-1}) of inorganic fertilizer. This is illustrated for wheat by Figure 5.14, where yields and fertilizer inputs are expressed as percentages of the 1959 values. It is apparent that the increase in phosphorus input has been similar to the increase in yield but that the rises in nitrogen and potassium inputs have been disproportionately rapid. This implies

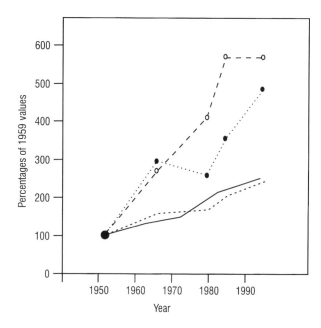

Fig. 5.14 Average annual United Kingdom wheat yields per unit area (solid line) and average annual inputs of phosphorus (broken line), nitrogen (dotted line with solid circles) and potassium (dotted line with open circles) per unit area. All expressed as percentages of their 1959 values. Derived from Royal Commission on Environmental Pollution (1996).

that for nitrogen and potassium at least, there is a considerable input of nutrient that is not getting into the crop and is likely to be finding its way, eventually, to the river system. Newman (1997) considered a phosphorus budget for a typical British wheat farm in the 1980s and concluded that, even for this element, there was a surplus of $0.9\,kg\,ha^{-1}\,year^{-1}$ available for loss by leaching $(0.4\,kg\,ha^{-1})$ and 'soil erosion by water' $(0.5\,kg\,ha^{-1})$. Other published information shows that a substantial percentage of the nitrogen, applied to crops other than grass, may find its way into watercourses (Table 5.7). These substances cause enrichment of streams and lakes with consequent effects on aquatic vegetation that include overgrowth within streams and encouragement of potentially toxic blue–green algae in some lakes.

The numbers of farm animals in Britain have increased considerably during recent decades. Numbers in 1994, expressed as percentages of numbers in 1940, were cattle 124%, sheep 188%, pigs 200% and poultry 202% (Royal Commission on Environmental Pollution, 1996). Events likely to cause organic enrichment of watercourses account for some 75% of reported farm pollution incidents in England and Wales (Fig. 5.15). There will be many other organic enrichments that are not detected, let alone reported. Organic materials such as milk wastes and spillages, cattle slurry, pig and chicken faeces and silage liquor can all cause substantial deoxygenation of recipient waters. Silage liquor has

Table 5.7   Application of nitrogen to various crops in England and Wales and the percentages available to be leached. Derived from Royal Commission on Environmental Pollution (1992b).

Crop	Used in farming ($kg\,ha^{-1}$)	Percentage available to be leached
Grass	119	5
Winter barley	144	20
Spring wheat	144	35
Oilseed rape	240	42
Sugar beet	122	65
Potatoes	200	56

about 300 times the oxygen demand of human sewage. These materials have an additional effect in that they all contribute to the suspended solids load. The threats from silage have reduced in recent years with the increasing tendency to store it in large round bales. Nevertheless, the Scottish Environmental Protection Agency still found it necessary in 1998 to remind farmers of the need to exercise extreme care in its production, storage and use. Serious problems from suspended solids and BOD can arise from slurry entering watercourses. Particular dangers arise from spreading slurry close to watercourses and/or on frozen ground. If such activity is followed by heavy rainfall, substantial amounts may be washed into the streams.

The transport of suspended solids by streams is normal but the amount of such material appears to be increasing and much of the increase stems from changing agricultural practices. Wind erosion of soil occurs especially when light soils in

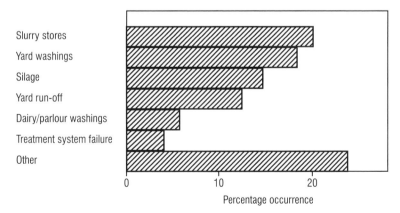

Fig. 5.15   Percentage distribution between causes of the 3147 farm pollution incidents reported in England and Wales in 1990. The figure separates the 'top six' that were responsible for 76% of the total. In Scotland in the same year, there was a lower percentage of incidents involving slurry (10.7%) and a higher percentage (52.5%) concerned with silage. Derived from Royal Commission on Environmental Pollution (1992a).

dry areas are turned from pasture into arable. A prime example was the ploughing of chalk grassland and alluvial grazing land that began in parts of southern England some thirty or more years ago. More recently, there has been a disturbing trend towards increasing water erosion of soil through overstocking with sheep and cattle, especially the former, in the more western parts of Britain. The effect becomes most apparent in areas where a mild climate allows stock to be overwintered in the fields and supported mainly by feeding with stored feed, especially silage. The combined effects of the stock grazing, trampling and standing about waiting for the 'meals on wheels' is to destroy the vegetation cover and 'puddle' the soil, thus giving a rapid run-off and a soft and readily eroded soil surface. The resultant mixture of soil and nutrient-rich material then finds its way to the streams by direct run-off and, where the management is particularly bad, by being deposited thickly on road surfaces by the vehicles used to deliver the daily feed ration and then washed off by subsequent rainfall into ditches and streams.

The final type of agricultural impact overlaps, to some extent, with those already described and can best be termed 'structural damage'. The most obvious examples concern physical damage to the stream that impairs its value as salmonid habitat (and indeed for wildlife in general). These may include insensitive abstraction of gravel to create roads or hard standings, and attempts to straighten and 'train' streams. These activities temporarily increase the suspended solids load downstream, may remove or damage spawning areas and can lead to instability of the banks and bed that may cause appreciable erosion over a period of years. Trampling by animals can initiate damage to stream banks and increased erosion rates (Figs 5.16 and 5.17). The effects of excessive animal trampling in damaging field soil structure and increasing erosion have already been noted.

As farming is the largest form of land use in Britain (Fig 5.1), it is hardly surprising that it has a major influence upon the aquatic environment. That influence is often more harmful than it need be and this arises from a complex of technical, social and political factors that tend to militate against us taking the holistic view that would help to set agriculture within a broader environmental framework. Four of the more important issues can be briefly summarized:

(1)   Modern farming is a technically advanced and complex enterprise. Many farmers are probably fully exercized in keeping abreast of technical advance within the farm and have little time to fully consider the environmental ramifications of their activities.

(2)   The true economics of agriculture in Britain (and the European Union) are obscured by a high level of subsidies in cash and in kind. Over 50% of total farm income in the United Kingdom during 1990 was subsidy (Royal Commission on Environmental Pollution, 1996). Some of these subsidies (e.g. headage payments on livestock) actually encourage overstocking.

Fig. 5.16 Damage to river banks as a result of trampling and grazing by cattle. Southwestern England, 1996. Photograph by the Environment Agency.

Even drastic reduction of the subsidies might not, however, be a solution. The operation of proper market forces might give greater 'economic efficiency' but this would not necessarily give greater 'biological efficiency' or sustainability. It is a truism that market forces are a useful tool for setting prices but that they are unable to assess costs.

Fig. 5.17 Damage to river banks as a result of sheep grazing and trampling. Northern England, 1996. Photograph by the Environment Agency.

(3)   There is often no adequate natural feedback mechanism because many of the environmental costs created on farms are actually borne elsewhere. This is particularly true of effects upon the aquatic environment. The average cost of lost output through soil erosion in 1985 was less than £100 per farm and the cost of reducing the erosion risk might well exceed this. The off-farm costs can, however, be much higher and are not always borne by the farmer. A very serious erosion event in Sussex (England) caused on-farm losses of about £13 000 and off-farm damage of about £420 000 (Royal Commission on Environmental Pollution, 1996).

(4)   The penalties imposed by law for pollution incidents are often too small to be an effective deterrent. Mills (1989) quoted fines varying from £50 to £1600 in 1986 for agricultural pollution. In 1997 the average fine imposed for all water pollution incidents in England and Wales was £4300 compared with an available maximum of £20 000 in magistrates' courts and unlimited amounts in higher courts.

It is important that any future restructuring of agriculture within Britain or the European Union should ensure that proper account is taken of all the environmental issues, including those that affect salmonid fish.

Chapter 6
Assessment, Conservation, Improvement and Rehabilitation of Salmonid Habitat

(a model environment)

Summary

River restoration works should, ideally, be catchment-based such as to enable the river ecosystem to recover as rapidly as possible via natural processes. In practice this ideal is rarely fully achieved. Restoration for salmonids is usually simply a part of this general restoration process, though various measures aimed specifically at rehabilitation or enhancement of salmonid habitat are also possible. There are often problems in assessing what a given system was like when pristine, and also in determining the trout and salmon population densities and structure that were typical of the pristine river/stream. During their execution, restoration works need a high-quality, multi-disciplinary scientific input. It is important that the aims of each project are carefully focused and that the work is well planned and supervised. There should be detailed and careful recording of methodology and costs. Pre- and post-restoration/enhancement monitoring work must be of high standard, take full account of spatial and temporal variations in fish numbers, and yield results capable of rigorous statistical analysis and testing. All details and results should be archived and scientific findings should be published in peer-review scientific journals.

We have substantial knowledge of the environmental requirements of trout and salmon. Despite this, our ability to define and quantify the mechanisms is limited. It is often inadequate to enable us to restore or conserve habitat in an effective and cost-efficient manner. There is, therefore, a need for further fundamental research of a quantitative and critical nature.

There are numerous publications and symposium volumes that describe the physical dynamics of stream/river systems and cover a variety of aspects of general river management and restoration; a selection is listed in Table 6.1.

Table 6.1 Some useful general works on stream and river dynamics, restoration and enhancement.

Authors	Date	Brief description of contents
Boon *et al.*	1992	Symposium on river conservation and management.
Brookes & Shields	1996a	Physical river channel restoration, with emphasis on fluvial dynamics.
Central Fisheries Board	1995a	Habitat improvement for salmon and trout in small streams – a layperson's guide.
Central Fisheries Board	1995b	Physical restoration of salmon and trout streams – a layperson's guide.
Cowx	1994a	Symposium on freshwater fisheries restoration.
Cowx & Welcomme	1998	EIFAC handbook on river fisheries restoration.
Environment Agency	1998b	Fish habitat improvement – a layperson's guide
Harper & Ferguson	1995	Symposium on river management.
Mills	1991	Symposium on salmon river rehabilitation.
National Research Council	1996	Conservation and restoration of Pacific Salmon.
Newson	1994	General work on hydrology and the river environment.
Petts	1984	Impoundment and river management.
Slaney & Zaldokas	1997	Procedures for fish habitat rehabilitation with special reference to British Columbia.
Taccogna & Munro	1995	Layperson's guide to stream and wetland assessment and restoration with special reference to salmonids.

Selected literature dealing with more specific topics is listed in Table 6.2. Mann & Winfield (1992) reviewed the literature on restoration of riverine fisheries habitats and noted that most of the best handbooks on techniques for restoration of salmonid habitat are based on North American experience, though not all of the techniques are appropriate for UK rivers. They drew particular attention to Parkinson & Slaney (1975), Maughan *et al.* (1978), Finnigan *et al.* (1980), Gore (1985), Hunt (1988) and Hunter (1991). It is not intended here to review this huge volume of literature in detail or to present a 'recipe book,' but rather to bring the material to the attention of the reader and then attempt to examine, in a little more detail, certain principles that appear particularly worthy of attention.

Table 6.2   Useful works on specific topics.

Authors	Date	Brief description of contents
Guidance notes		
Forestry Commission	1993	Forestry and the aquatic environment.
National Rivers Authority	1994b	Protection of the aquatic environment through development plans.
Environment Agency	1997b	Buffer zones.
National Rivers Authority	1998d	River bank erosion and conservation.
Environment Agency	1998c	Diffuse agricultural pollution.
Ministry of Agriculture, Fisheries and Food & Welsh Office Agriculture Department	1998a	Agriculture and the soil.
Ministry of Agriculture, Fisheries and Food & Welsh Office Agriculture Department	1998b	Agriculture and the water.
Giles & Associates	1998	Fisheries and wildlife conservation.
Fish passes, turbines, screens, counters		
Scottish Office Agriculture and Fisheries Department	1995	Guidance on fish passes and screens.
Beach	1984	Fish pass design criteria.
Carling & Dobson	1992	Fish pass design and evaluation.
Clay	1961	Design of fishways.
Davies	1988	Fish passage and turbines.
Gregory	1989	Water schemes and fisheries.
Holden	1988	Automatic fish counters.
Turnpenny	1989	Exclusion of fish from intakes.
Winstone *et al.*	1985	Weirs and salmonids.
Mann & Aprahamian	1996	Fish passes.
Salmon Advisory Committee	1997	Fish passes and screens.
Gravel cleaning		
Kondolf *et al.*	1987	Flushing flow requirements for brown trout.
Mundie & Crabtree	1997	Effects of gravel cleaning in a salmonid spawning channel.
Scott & Beaumont	1994	Gravel cleaning in chalk streams.
Stocking		
Egglishaw *et al.*	1984	Stocking streams with salmon eggs and fry.

6.1   Definitions and background relating to river restoration

Stream/river restoration may have one or more of a number of objectives; these objectives must be identified at the outset. The main interest in this book is in restoration of streams/rivers as habitat for trout and salmon. It is not, however, possible to consider this narrow objective in isolation. Rather, there is a need to restore the structure and functions of the river ecosystem (including its catchment, as far as is practicable) but perhaps with particular reference to trout and salmon. Brookes & Shields (1996) quote four definitions of river restoration. These range from simple statements on water quality restoration (Herricks & Osborne, 1985) or channel restoration (Osborne *et al.*, 1993), to the more comprehensive view (Cairns, 1991) that restoration is 'the complete structural and functional return to a pre-disturbance state'. Gore (1985) further incorporates the concept of enablement. Perhaps a more realistic definition of restoration is 'the return of an eco-system to a close approximation of its condition prior to disturbance' (National Research Council, 1992). Brookes & Shields (1996) recognize three different activities related to stream improvement. 'Full restoration' is complete structural and functional return to a pre-disturbance state, and may be achieved by natural recovery, enhanced recovery or direct intervention. 'Rehabilitation' is partial return to pre-disturbance structure and function, and is achieved either by enhanced recovery or direct intervention. 'Enhancement' is any improvement in environmental quality, usually via direct intervention.

From this welter of terms and definitions, several characteristics of good restoration work become apparent. First, restoration is, ideally, an enabling process that creates a suitable framework within which the system can recover via natural processes. Second, the results of good restoration are sustainable with little or no need for subsequent intervention (Gregory & Bisson, 1997). Third, it is necessary to restore both the structure and the functions of the system and this is best done at a catchment level.

In practice it is often difficult to define the 'pre-disturbance' state. In many regions, including much of western Europe, the landscape and watercourses have been modified by man over a number of centuries and many of the changes are irreversible. Most systems are subject to multiple uses that may include fisheries, navigation, wildlife conservation, recreation and others on the watercourses, and a wide variety of land uses within the catchment. In such circumstances, 'by far the most common tendency is to attempt to live with modifed systems through a series of interventions and management strategies designed to lessen the impact of stress' (Cowx & Welcomme, 1998). Most of these restorations are therefore imperfect, and can at best be described as 'rehabilitation' or 'mitigation'. Some of these matters are considered in more detail later in this chapter.

A stream/river is a dynamic system that cannot be viewed realistically in isolation from its floodplain or its catchment. As the water in a stream flows downhill the potential energy of altitude is converted into the kinetic energy of motion. The kinetic energy of the water tends to increase with distance down-stream. Excess kinetic energy is usually dispersed as sound, as turbulence and via friction with the bed and banks. Larger quantities of water are carried during spates and water in excess of the capacity of the channel flows out into the floodchannel or floodplain. This reduces the rate of run-off. Much larger quantities of kinetic energy are present during spates. Some of the extra energy gives higher water velocities, greater turbulence and more noise but much of it may be dispersed through erosion of bed and banks and by the transport of bedload (Chapter 3). As a consequence, the sinuosities of the channel move downstream and may become accentuated. The riffle-pool sequence also moves downstream in the course of time. Interference with this dynamic equilibrium occurs when changes in the catchment modify the run-off pattern, when channels are straightened and/or smoothed, and when floodplains are restricted. Such interference generates a need for physical restoration works. If these works are to be more than glorified landscape gardening they need to restore the function as well as the form of the system. This requires inputs of expertise in both physical science (hydrology, hydraulics, geomorphology) and biological science (biology, ecology). Neither of these two branches of science is likely to produce good results when used in isolation from the other.

6.2   General principles for fishery rehabilitation

Cowx (1994b) set out a general scheme of implementations and feedback mechanisms for rehabilitating inland fisheries. The feedback mechanisms consist mainly of questions about whether or not the proposed methods are acceptable and/or efficacious. Shorn of these, the scheme consists of the following steps:

(1)   to evaluate the fishery
(2)   to identify the problem areas
(3)   to propose methods to improve the fishery
(4)   to implement the proposals (if acceptable and practicable)
(5)   to monitor the results.

The last stage leads to two outcomes. First, a re-evaluation of the fishery to see whether or not it has been improved and also as a guide to possible further improvements. Second, dissemination of the results. Variants of these themes will recur throughout this chapter.

6.3   Assessment of habitat in terms of the need, feasibility and outcome of restoration

Ideally, this assessment should include the whole catchment, even though the restoration may include only a part of the catchment or stream. The best basis for planning is a set of detailed physical, chemical and biological surveys (Whelan, 1991). General principles for the planning and management of restoration projects have been outlined by Rundquist *et al.* (1986) and Cairns (1990). The latter gives a useful checklist of 22 questions to be addressed before the work begins. These general principles apply to restoration work for fish and other aquatic biota and were summarised by Mann & Winfield (1992) as four points.

(1)   What is the aim of the restoration?
(2)   Which environmental attributes need to be restored?
(3)   What are the environmental specifications for an alternative restoration scheme?
(4)   Would natural processes lead to restoration more effectively?

The first of these questions may present difficulty. If the aim is to return the environment to its original condition, then adequate information about the original condition will be required but may not be available. An affirmative answer to the fourth question leads to the 'do nothing' option and this should always be borne in mind.

An appraisal of salmonid stocks is a major element in the assessment of salmonid habitat for restoration or improvement and a brief general discussion of this was given by Cowx (1995).

A stock-recruitment curve (Chapter 2, and Fig. 2.7) is a valuable tool for use in assessing salmonid populations. In populations where, other than during occasional exceptional years, the points fall on or near the flat part of the graph, there will be no value in any management measures aimed simply at increasing the input of eggs or fry. This is because mortality rate will rise with numbers and the population will still be limited to the carrying capacity. In such circumstances the most useful strategy would be:

(1)   To make efforts to conserve the existing habitat and to prevent over-exploitation of the adults.
(2)   To explore the possibilities for instream improvements aimed at increasing the quantity and/or quality of habitat so as to increase the carrying capacity (i.e. raise the line b–d on Fig. 2.7 to a higher level).

It is worth noting that even in the studies by Buck & Hay (1984), Gardiner & Shackley (1991) and Elliott (1994) on populations largely regulated by density-dependent mortality, approximately 20% of the data points fell on the ascending part (a–b) of the graph. It might be worth investigating the causes of this and exploring the possibilities for mitigation. In those populations where all or most of the data points lie on the ascending (density-independent) part of the graph, the numbers of recruits are usually at population densities below carrying capacity. The mechanisms postulated as causes of this density-independent limitation include over-exploitation of adults (Watt & Penney, 1980; Chadwick, 1985), washout of intragravel stages (Crisp, 1993a) and/or restriction of adult access by obstructions (Crisp & Beaumont, 1995). In these populations the restoration/improvement strategy depends upon accurate identification of the limiting factors and this is not usually a simple matter. The next step is to attempt to eliminate or ease the limitations to bring the points towards line b–d. Dependent upon the precise problem, this could include one or more of measures such as:

(1)   Ensuring a larger escapement of spawners by reducing the exploitation rate.
(2)   Removal of physical obstructions to allow better access for potential spawners.
(3)   Improvement of the hydrological and/or suspended solids regime by blocking drains and/or encouragement of better catchment management in order to reduce the probability of washout or siltation of redds.
(4)   Stocking with hatchery reared fry or parr.

Should it be possible by these means to regain the flat (density-dependent) part of the graph, the measures listed in the previous paragraph could then be used to attempt to increase carrying capacity. It is worth noting, however, that regulation mainly by density-independent mortality is typical of populations close to the edge of their range of tolerance, and in such populations the main limitation may be a natural one imposed by climate and therefore not amenable to human control.

   A stock-recruitment curve can shed valuable light on which measures for enhancement or restoration are most likely to be appropriate. Unfortunately, the production of such graphs requires considerable research effort and we only have them for a limited number of mainly small streams. A possible way forward for whole river systems was explored by Elliott (1992) with reference to sea trout rivers in a selection of regions in England and Wales. It was assumed that year-by-year variations in recorded catches by commercial and/or rod fishing were a useful index of year-upon-year variations in recruitment. The coefficient of temporal variation (CV_t) was expressed as a percentage and used as a measure of variation between years in each river or river system. CV_t% was calculated as $100s/x$, when x is the mean catch over a number of years and s is the standard deviation of that estimate. Elliott concluded that rivers with high between-years variability in catch (CV_t% > 86%) usually provided low catches and had populations regulated chiefly by density-independent factors, whilst rivers with low values of CV_t% (< 50%) usually gave medium or high catches and had populations regulated chiefly by density-dependent factors. It is important to note that this methodology depends upon assumptions that are not, at present, capable of objective proof and that the conclusions reached must, therefore, be tentative. Nevertheless, this approach is potentially useful and it would be interesting to see it extended to other UK rivers and applied also to salmon catches.

In the absence of a stock-recruitment curve, other methods of stock assessment must be used. These may include analyses of anglers' catches (Gee & Milner, 1980), electrofishing surveys or mark-recapture exercises (Seber, 1973), automatic fish counters (Holden, 1988), netting and redd counting. A need for restoration work may be indicated in either of two ways. First, if, within the river being considered, some index of population numbers has declined more than would be expected from past records for the site. Second, if, within the river being considered, some index of population has fallen below the level to be found in similar rivers that are in a relatively pristine state. The difficulty in using either of these criteria is that to demonstrate the existence of a real problem or change, the observed trend needs to exceed the sort of fluctuations in time or space that might be expected to occur naturally and this difference needs to be amenable to statistical testing.

Bohlin *et al.* (1989) referred to the precision of electrofishing estimates and they defined the precision of the estimates in terms of guideline values of coefficient of variation (CV%) such that class 1 precision gives a CV% of <5% and classes 2 and 3 give values of <10% and <20%, respectively. Cowx (1995) followed the same line and tabulated the suggested precision levels required for various management activities. Thus the evaluation of the status of fish stocks requires a relative or absolute estimate with a precision of class 2 or 3, whereas an environmental impact assessment will require precision of class 1 or 2. Sutcliffe (1979) defined 'precision' (not to be confused with 'accuracy') as a measure of the reproducibility of a method, measured by repeated observations on the same sample. By this token, the precision classes of Bohlin *et al.* can be taken to refer to the theoretical reproducibility of results obtained from a succession of estimates made upon the same population within the same sampling site by the same team, and its coefficient of variation could be termed the coefficient of variation of the method (CV_m). In practice, it probably refers to repeated use of the same team and equipment in a series of similar reaches in the same stream at a point in time and may, therefore, also incorporate part of the coefficient of spatial variation (CV_s). In any event, the values of CV quoted by Bohlin *et al.* are very low when compared with the coefficient of variation with time (CV_t) that can be seen in many salmonid populations as a result of natural temporal fluctuations (Elliott, 1992 and Table 6.3). Even in populations that are usually regulated by density-dependent mortality, the ratio of highest to lowest observed values may be as high as 2 or 3 to 1 (Chapter 2). It is therefore evident that there is considerable between-years variation in population numbers in salmonids and this creates major problems, initially in establishing the need for restoration work, and later in assessing its efficacy.

Winstone (1989) used published information from several different methods of salmonid population assessment in an instructive exercise that should, perhaps, be required reading for all engaged in stream restoration or enhancement work. It is apparent (Table 6.4) that all of the assessment methods give relatively high

Table 6.3 Summary of declared rod catches of sea trout and salmon in the Welsh Region of the National Rivers Authority/Environment Agency over a period of ten years (National Rivers Authority 1993, 1994c, 1994d, 1995; Environment Agency, 1997c) and of electrofishing estimates of the population of 0-group trout in a section of a small sea trout nursery stream in Wales (from Crisp & Beaumont, 1995; plus unpublished data) over a period of 14 years. For each data set the range of observed values is given, together with the ratio of the highest to lowest values and the calculated value of $CV_t\%$.

	Sea trout rod catch, Wales	Salmon rod catch, Wales	0-group trout, Afon Cwm, Wales
Period	1986–1995	1986–1995	1984–1997
Range of values	3355–15 989	4634–35 727	7.2–156.0
Highest:lowest	4.8:1	7.7:1	21.7:1
$CV_t\%$	52.3	55.7	76.3

values of $CV_t\%$ and that electrofishing for 0-group trout and salmon probably gives the lowest values. From the calculated values of CV_t, Winstone estimated the duration of pre- and post-intervention monitoring needed to detect a given fractional reduction in population as a result of the intervention. The example quoted was the proposed construction of an estuarine barrage that was likely to cause a decrease in population, but the same arguments and statistical logic would apply to the detection of changes arising from habitat deterioration (to demonstrate the need for restoration/improvement) or from habitat improvement (to demonstrate the effectiveness of the work). Figures 6.1 and 6.2 are redrawn from Winstone's figures for the stages that were shown to have the lowest values of CV_t (Table 6.4) and it is clear, even for these stages, that data are needed from a substantial number of years, both before and after intervention, if the minimum detectable fractional change is to be reduced to a meaningful level. Bisson *et al.* (1997) also noted the need for long data runs and argued that this implies a need for planning horizons to extend well beyond the normal scope of long-term plans.

Table 6.4 Five methods of population assessment together with ranges of observed values of $CV_t\%$ from various published long term (five or more years) studies. Derived from Winstone (1989).

Stage	Method	Details	$CV_t\%$
Juvenile	Electrofishing	0-group salmon	12–28
		0-group trout	19–20
		>0-group trout	24–28
Smolts	Trapping	–	18–45
Adults	Rod catch	–	30–54
Adults	Trapping	–	24–42
Redds	Redd count	–	34–54

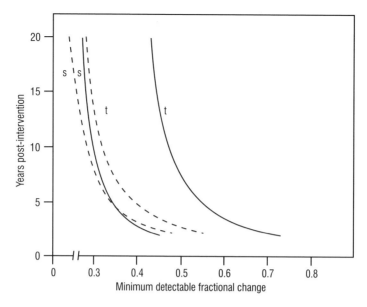

Fig. 6.1 Minimum fractional change that can be detected for 0-group salmon(s) and trout(t), relative to the number of years of post-intervention data, when pre-intervention data are available for five years (solid line) and ten years (broken line), redrawn from Winstone (1989).

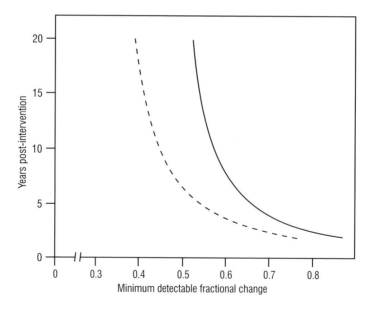

Fig. 6.2 Minimum fractional change that can be detected for >0-group trout, relative to the number of years of post-intervention data, when pre-intervention data are available for five years (solid line) and ten years (broken line), redrawn from Winstone (1989).

Comparison of the existing situation in a river with that expected in a river in an ideal or pristine state is a further method of assessing the need for restoration. Very often, the 'ideal' is in the imagination of the would-be restorer and has little or no objective, let alone quantitative, basis. A rather more objective approach can be made via mathematical models.

A hydro-ecological model known as Instream Flow Incremental Methodology (IFIM) using the Physical Habitat Simulation' (PHABSIM) system (Bovee, 1982) was developed in North America and is now being used in Europe. An introduction to its use in the UK is given by Gustard & Elliott (1997). The method assumes that for a given species and life stage, a Habitat Suitability Index (HSI) can be based on four variables – water depth, mean water column velocity, substratum and cover. These may be assessed by field observation, by study of literature or by expert opinion. For each life stage, it is then possible to assess the Weighted Usable Area (WUA). This methodology has been criticised by Heggenes (1988) and Elliott (1994). The latter drew attention to three potential problems:

(1) Between-years variations are ignored and these can be large, as shown above.
(2) Linear or log-linear relationships are assumed, often erroneously.
(3) Interactions between habitat variables are not taken into account.

Heggenes gave particular stress to point (3) and quoted De Graaf & Bain (1986) as evidence of such interactions for juvenile salmon. The very limited number of habitat variables used in this approach is also noteworthy. With reference to point (2), the sorts of relationships indicated in Figs 4.14, 4.15 and 4.16 above are too complex to be represented by simple linear models; in addition, fish size would need to be incorporated as an additional variable. Mundie & Bell-Irving (1986) noted the failure of IFIM to recognize the importance of dominant flows in maintaining the physical characteristics of rivers. This point has particular relevance to spate rivers containing gravel that serve as spawning and nursery areas for salmonids. Further criticisms relate to the scale of assessment, and particularly to the fact that hydraulic differences between riffles, pools and dead water are largely ignored (Padmore, 1998). The most important drawbacks of this methodology are that it is very simplistic and, despite initial appearances, it has an appreciable subjective content, especially with regard to the assessment of substratum and cover – even when gravel composition is analysed quantitatively (Chapter 4). The subjective content is increased further when there is recourse to expert opinion. There is scope for further development of this methodology (e.g. Capra *et al.*, 1995; Castleberry *et al.*, 1996) and it would be pleasant if this could be achieved with minimal additions to the stock of jargon and acronyms already generated by this theme! 'Habscore' was originally developed (Milner *et al.*, 1985) as a method of evaluating the salmonid habitat quality of streams and rivers in Wales. It has been developed subsequently as a computer software spreadsheet accompanied by a field manual and a software manual, and a later version is applicable to UK streams outside Wales. It was originally based on multiple regression analysis for 166 sites that were examined over a period of three years. The modelling assumptions were:

(1)   the sites were a random selection
(2)   the sites were representative of the types of sites to which the model would be applied
(3)   the sites were 'pristine', i.e. without any water quality problems likely to adversely affect the fish populations
(4)   recruitment was not limiting.

A large number of catchment and site variables are measured, including the catchment area, gradient, water conductivity, position of the site relative to the river source and mouth, and the length, depth, slope, flow, substratum, cover and vegetation of the study section. The output is a habitat score for the section which is then compared with the observed fish population to give a measure of habitat utilization. This can be taken as an index of the extent to which the site departs from the ideal, and hence of the scope for improvement or restoration. It can also be used on a 'before and after' basis to test the effectiveness of improvement work. The authors take pains to explain that this model is simplistic and its application should not, therefore, be pushed too far. It is, nevertheless, less simplistic than IFIM, though it shares some of the problems of that model. Particular problems are:

(1)   The assumption that the streams used in the original surveys were all 'pristine'; any errors here would lead to underestimation of carrying capacity.
(2)   The assumption that in the original surveys recruitment was not limiting and that 0-group numbers were at carrying capacity; this overlooks the large amounts of between-years variation that can be observed and will, again, tend to give underestimates of carrying capacity.
(3)   The possible introduction of an element of subjectivity into the assessment of such site variables as substratum, cover and vegetation.

Despite these limitations, models of this type, especially if developed further, can be a valuable starting point for habitat assessment, provided the results are used with caution.

6.4   Conservation, improvement and restoration of salmonid habitat

These three issues are best considered together because they really represent stages in a continuum. 'Conservation' largely consists of recognizing good salmonid habitat and taking steps to retain or enhance it; 'improvement' consists of recognizing inferior salmonid habitat and seeking to make it better; and 'restoration' includes the recognition of what was once moderate-to-good habitat that has been degraded and seeking to make it good again. It is convenient to emphasize the improvement and restoration aspect, whilst bearing in mind that conservation can be taken broadly as the process of ensuring that we do not create any further sites in need of restoration or improvement. Most of the methodology for these processes is implicit in Chapters 4 and 5. The broad issues can be considered under four headings: fishery management, water quantity and quality, obstruction of fish movement, and physical habitat degradation. Each will be considered in turn.

6.4.1   Fishery management

This relates to issues such as exploitation and regulation of fisheries, together with any possible impacts from stocking and fish farming. The major aspects have been covered in Chapter 5 and earlier in this chapter and will not be repeated here.

6.4.2   Water quantity and quality

These can be considered under four main headings: chemical pollution, suspended solids, temperature effects, and modified flow regime.

Chemical pollution can arise from both organic and inorganic poisons, and also from the creation of oxygen demand by the breakdown of various organic materials. All of these materials can be a problem, even when not applied directly to the stream, because they may be washed into watercourses from other parts of the catchment. The obvious remedy is to seek to cut off pollution at source by the use, as a last resort, of existing legislation, adequate enforcement and the imposition of appropriate penalties on offenders. Difficulties often arise in tracing the source or sources especially when intermittent and/or diffuse. The scale of the potential problem is indicated by a survey of 100 silage storage facilities in Scotland. This showed that all of them had faults leading to some degree of leakage (Department of Agriculture and Fisheries for Scotland, 1982a) though the risks involved have been reduced in recent years owing to an increase in the use of large round bales for silage storage. As the number and potency of the chemicals available (particularly to agri-business) continues to increase, it is important to ensure that the advice and legislation keep pace with technological developments. A good example is the development of new forms of sheep dip and a consequent increase in their capacity to damage the aquatic environment (Chapter 5).

It would appear that within the UK the most important influence of coniferous forests upon salmonids is the effect of acidification and monomeric aluminium in areas with poorly buffered soils. This alone may explain the absence or scarcity of salmonids in some afforested streams. Evidence from mid-Wales (Chapter 5) suggests that clear felling may actually cause a further deterioration of water quality, though the duration of this effect is not yet known. There are various liming techniques that can alleviate the acidification problem but these are expensive, may have harmful effects on other components of the catchment biota (e.g. some species of *Sphagnum* moss), and may need fairly frequent repetition of the treatment. In the absence of acidification effects, forestry activities may still influence salmonids via changed patterns of flow, bed scour, sediment transport and silt deposition. Hartman *et al.* (1994) commented 'There is a high level of complexity in natural ecosystems. We cannot always judge correctly what the critical factors or times are in the life of populations of fish... There may also be sets of physical processes that overlap and compound their effects. Forestry impacts have different effects on different species and life stages of fish.' Similar comments were made by Crisp & Beaumont (1998) with regard to our inadequate understanding of the details of the processes in which we may seek to intervene. Some of the effects of past forestry activity may be irreversible and restoration is always a poor substitute for habitat protection.

We have already noted that concentrations of suspended solids in streams can be increased by soil erosion, road run-off and various disturbances caused by civil engineering, agriculture and extractive industries. It is far from clear whether or not existing legislation can cope with 'pollution' by inert materials whose harmful effects may be separated in both time and space from their origin. A prime example is the washing into a stream by heavy rainfall of field soil deposited in its catchment at some earlier date as a result either of windblow or of wet deposition on roads. This material might cause immediate damage to fish via abrasion of their gills and, at some later date and further downstream, it might cause further damage by choking redds. Further complications arise in detection for two reasons. First, it is not possible to measure easily suspended solids concentration in the field. Second, it is difficult to be certain how a given concentration relates to the 'normal' value in any given stream because concentrations vary considerably with discharge (Fig. 3.1). Accumulations of silt in spawning gravels can be removed by various washing, raking and riddling techniques (Solomon, 1983; Kondolf *et al.*, 1987; Scott & Beaumont, 1994; Mundie & Crabtree, 1997). This cleaning and decompaction of the gravel is expensive, but it can improve gravel quality for salmonid spawning. Two important points should be noted. First, the sediment washed out will settle further downstream and may cause damage to spawning gravels there. Second, recompaction and re-infiltration with sediment will occur after cleaning. Where there is a good source of sediment, refilling of the gravel interstices may be rapid (Carling & McCahon, 1987). Gravel cleaning is likely to be a long lasting solution only if at the same time there is a reduction of availability of sediments from the catchment to enter the stream. This requires a detailed investigation of the sources and concerted action to reduce the inputs.

Changes in temperature regime can have a wide variety of effects on all the life stages of salmonids. Heated effluents from industry can be cooled by the use of cooling lagoons or cooling towers, dependent upon the amount of heated effluent produced. It is worth noting however that the materials used to treat the timbers used in some cooling towers can cause toxic effects. Where impoundments stratify in summer and have multiple draw-off levels, then (as noted in Chapter 5) judicious choice of combinations of draw-off level and discharge can be used to give a water mix that simulates or even improves upon the natural summer temperature regime of the river.

Modifications of flow regime do not only have direct biological consequences but also have indirect effects via wetted area, water depth, water velocity and the movement of sediments and bedload (Fig. 5.3). Where there is direct human control over flow regime (e.g. at impoundments) it is important that the compensation flow and the fluctuations in flow are as ecologically sensitive as is possible within the constraints of the engineering function of the structure. A good example of such sensitive management is the gradual increasing or decreasing of discharges at, for example, hydroelectric facilities, rather than the relatively abrupt changes from high to low discharge and back again that were once practised.

The creation of open drains on upland catchments has, in the past, been a cost to the taxpayer in subsidy and has led to further costs in the form of damaged salmonid habitat. It is heartening that there is now an increasing tendency amongst owners and managers of the uplands to realize that this exercise was of very limited value to the landowner and was also harmful to the aquatic environment. The present trend is to allow these drains to become obstructed by natural causes or to actively block them.

6.4.3   *Fish movements*

Both anadromous and resident salmonids need to move between different parts of the river system to reach spawning areas. This movement may be hindered or prevented by a wide range of natural and man-made obstructions. The latter include weirs and impoundments. Where weirs are redundant it may be possible to breach or remove them. Otherwise, both weirs and impoundments can be bypassed by a variety of types of fish passes and lifts that have been described in the literature (Table 6.2). It is useful to assess the ability of salmonids of given sizes to pass given obstructions. Where the obstruction is surmounted by high-speed swimming, passage may be limited by some combination of water depth and water velocity on the face or crest of the weir and, knowledge of the water depths and velocities at different river discharges can be used, together with knowledge of fish sizes, to make intial predictions. It is also useful to take account of water temperatures (Figs 4.12 and 4.13). Winstone *et al.* (1985) made such a study of two weirs on the River Afan (Wales) and were able to estimate the approximate number of days in the year on which sea trout of different size ranges would be able to pass each weir. Such analyses can form the basis for more detailed studies and for possible remedial action.

6.4.4   Rehabilitation or enhancement of physically degraded habitat

The rehabilitation of physically degraded habitat is a major element in most projects. There are numerous published descriptions and diagrams of instream structures and practical guidelines for physical works (Tables 6.1 and 6.2). The present account will cover these matters in broad outline, together with some discussion of wider issues. The main techniques can be grouped under four headings:

(1)   Cleaning, restoration and construction of gravel beds for use in spawning.
(2)   Provision of exclusion/buffer zones to protect river channels and banks from animals and/or to mitigate some of the effects of coniferous forest; additional measures to stabilise damaged banks may also be needed.
(3)   Restoration of natural channel characteristics, primarily in terms of stream width, meanders and the riffle and pool sequence.
(4)   Miscellaneous instream structures intended either to aid in point (3) or to provide additional cover for fish and hence increase the carrying capacity.

The variety of methods for reducing the compaction and fines content of gravels has already been mentioned. Where gravel beds have been lost, it is possible to create new beds for spawning and incubation by use of suitable gravel mixes (Chapter 4) but two points must be considered. First, a natural stream is a dynamic system and natural gravel beds move inexorably downstream. In spate streams this may lead to a need to regularly replenish the gravel bed or to attempt to stabilize it in some way. Second, the value of a gravel bed for spawning and incubation depends upon the composition of the gravel and also upon the hydraulic conditions (especially intragravel flow) within the bed (Chapter 4). The scope for gravel restoration in the United Kingdom is probably limited, but may be significant (Solomon, 1983). The greatest scope is probably for gravel loosening and cleaning, especially in chalk streams (Solomon & Templeton, 1976; Kemp, 1986).

'Buffer strips' or 'exclusion zones' have been recommended for two main purposes. Farm stock can damage riverbanks by overgrazing the streamside vegetation and through physical damage by trampling. This damage is most marked where there is overstocking. This increases soil erosion rates and leads to the river channel becoming wider and, consequently, to reduced water depth and velocity. A proper solution to these problems requires a reduction in stocking density. Even in the presence of high stock densities however, the damage can be reduced by fencing off a strip of land adjacent to the riverbank and allowing animal access for drinking at only one or more specified points at which gravel ramps or similar hard structures are provided. There are grants available for such work and details are given in National Rivers Authority (1998). The Forestry Commission (1993) recommends the use of 'protective strips' or 'buffer zones' adjacent to streams in afforestation schemes. Five main reasons for this have been given in the past:

(1) To act as 'seepage zones' that reduce the rate of run-off into streams and the amount of suspended solids washed into them; some blocking of drains may also be desirable (Drakeford, 1981).
(2) To protect stream banks from erosion.
(3) To give intermittent shade.
(4) To give protective cover to aquatic biota.
(5) To improve allochthonous inputs to the stream from leaf fall and terrestrial (invertebrate) casualties (useful fish food and dissolved nutrients).

A strip with a minimum width of 5 m either side of the stream is recommended for channels of up to 1 m wide, and minima of 10 m and 20 m are recommended for channels of 1–2 m and over 2 m wide, respectively. These zones can be planted with predominantly light foliaged deciduous trees but there should be adequate openings for sunlight penetration. In practice, if left alone, such strips will usually be colonized naturally by trees such as willows. The intentions expressed in these guidelines are laudable and the measures proposed are likely to be helpful but there does not yet appear to have been any objective, quantitative study of their effects upon salmonids. The guidelines also note that 'reduction in the emissions of acid pollutants is the only way of solving the general problem of surface acidification'.

Many streams in both upland and lowland areas suffer the effects of physical degradation. The commonest forms are channel straightening and canalization, channel widening arising from bank erosion or direct human action, and loss of the pool-riffle series that is typical of natural gravel-bedded streams and rivers. Channel straightening often leads to loss of the pool-riffle sequence. It also tends to destabilize the bed and banks, and in spate streams this can lead to appreciable erosion and bed movement. These consequences may then give rise to a downward spiral in which attempts are made to stabilize the system by means of 'hard engineering' projects that may further exacerbate the situation.

Hartman *et al.* (1994) commented that a large number of manuals for instream enhancement structures exists but only one of them (Lowe, 1996) provides dimensions, rock sizes and other engineering specifications. They also point out that the instability of some high-energy (flashy) streams may be so high that frequent repairs may be needed to maintain the efficiency of some instream structures. What is becoming clear is that restoration of the physical structure of streams, especially those with low base flow indices, cannot readily be achieved by following 'recipe books'. The measures need to be planned and executed carefully for each individual catchment. They also need to take full account of the dynamics of fluvial systems (Brookes & Shields, 1996) and of the long timescale of certain fluvial processes (Petts, 1984) and should, ideally, be planned and overseen by a multidisciplinary team containing quality expertise in biology, fluvial processes and engineering. Often, having in some way removed or reduced any harmful activities by humans and animals, the ideal form of restoration for a stream, especially an unstable one, is to create a corridor in which it will have enough room to reorganize itself and then leave nature to take its course. Any necessary interventions, such as a need to prevent the corridor being overgrown by scrub, should be kept to a minimum. Hartman *et al.* (1994) suggest that the width of such a channel meander zone should be determined by the characteristics of the river. They used the facts that the radius of a stream meander is typically 2.3 times the bankful width, and that riffle-pool and meander bend sequences reoccur in stream lengths equivalent to five to seven channel widths, to infer that a typical corridor should be at least six channel widths wide. In unstable streams, more than twelve channel widths may be needed.

In addition to their use in helping to enhance the natural physical character-istics of the stream, instream structures can be used to increase the diversity of habitat and to provide cover for fish, with the aim of increasing carrying capacity. A large range of structures is possible, ranging from simple do-it-yourself methods of improving cover to more elaborate engineering projects (Table 6.1); there is evidence that they can be effective in increasing salmonid numbers or biomass (Glover, 1986; Naslund, 1987; O'Grady *et al.*, 1991). Two aspects of this belong to the category of things that are self-evident but nevertheless need to be said. The first is that the objectives and possible consequences of these instream 'improvements' need to be considered very carefully. In particular, consideration should be given to the effects on all of the salmonid life stages and species. For example, the creation of deep pools and good cover for large fish in what was originally a good spawning and nursery area is not only likely to destroy some spawning and nursery ground but will also increase the population of larger salmonids available to prey upon the juveniles. The second aspect is a related one, and refers to the interpretation of the results of these 'improvements'. It is, for example, possible to create deep pools and good hiding places for larger resident salmonids and to find that they are colonized speedily by such fish. The inference to be drawn from this is not that the work has instantly increased the numbers of these large fish in the water (though this may possibly be a longer term consequence) but simply that larger fish have moved in from other parts of the system. Thus restoration should not be a piecemeal process but should involve whole streams or, ideally, catchments.

6.5   Education and communication

Much of the damage done to salmonid habitat is a result of ignorance rather than malice. Considerable progress in conservation and restoration would probably be achieved by a very effective programme of information and education of some of the most relevant groups. These include farmers, foresters, civil engineers and officials in local and national government.

Most foresters have now recognized the possible impacts of their activities on salmonid fishes and appropriate guidelines have been produced (Forestry Commission, 1993) in a brief and simple format. These will, no doubt, be updated as our knowledge of the underlying mechanisms improves. The vital need is to ensure that the guidelines have been read and understood by forestry workers and sub-contractors in the field, at all levels of the heirarchy.

The Environment Agency (formerly National Rivers Authority) in England and Wales produces an increasing range of guidance literature. Examples include guidance to local planners (National Rivers Authority, 1994b) and notes for landowners and managers on river bank erosion (Environment Agency, 1998d) and land use (Environment Agency, 1999). It is important to ensure that such material is based on sound science, and is clear, good quality, disseminated as widely as possible amongst target groups, and updated regularly.

Civil engineers are becoming much more environmentally aware, especially as a result of the increasing requirement for sound environmental impact studies before some schemes are approved. The quality of the fisheries consultancy employed on such work varies from good to mediocre, and this can sometimes create problems. Perhaps (as in forestry and agriculture) the most crucial point that needs to be addressed is that of adequate on-site communication, particularly on large projects. It is of little value for the resident engineer to understand the possible harmful environmental impacts of a scheme if the people driving the diggers do not. It is, after all, the latter who actually have the physical capability to do the damage. One solution to this on larger projects can be the appointment of a competent ecologist to make regular site visits to advise the senior engineers on possible problems and to ensure that the necessary information actually passes right down through the system of line management.

In recent years there have been several valuable initiatives by statutory bodies such as the Environment Agency in England and Wales, the Scottish Environmental Pollution Agency and organizations such as the Farming and Wildlife Advisory Group, to highlight some of the detrimental effects of intensive agriculture on rivers and to offer guidelines for good practice. Guideline codes for farmers have existed for some years and have recently been updated by the Ministry of Agriculture, Fisheries and Food and the Welsh Office Agriculture Department (1998a,b). Uptake of these codes has, however, been slow. A study in 1994–5 showed that only 46% of farmers were aware of these codes, only 18% owned a copy of the water code, and only 5% owned a copy of the soil code.

Perhaps, for some of the reasons outlined at the end of Chapter 5, farmers do not always realize the full effects of their activities on the environment (including fish populations). There is clearly a need for a substantial programme of information, education and advice. One important point to be communicated is that good practice, in some circumstances, is a paying proposition.

6.6   Design, monitoring, recording and dissemination

The National Rivers Authority (NRA) inherited various improvement and restoration schemes from its predecessor organizations and, subsequently, started others. As part of an audit of progress and performance, an independent review was comissioned of 44 projects in England and Wales of which 52% were the responsibility of the NRA and its predecessors, 20% were the responsibility of riparian owners, and the remainder were under joint sponsorship (Mann & Winfield, 1992). This report highlighted a number of weaknesses in the monitoring and reporting of restoration schemes in England and Wales. At much the same time, Hartman *et al.* (1994) made an assessment of the results of a long-term (> 20 years) study of a salmonid stream in British Columbia (Canada). This assessment was made with a view to gaining insight into the types of processes that should be taken into account in improvement and restoration projects. There is a striking similarity between these two studies in the types of generic weaknesses noted in the conduct of current projects; the main points are summarized below.

6.6.1   Science base

Both analyses found that the science base of many projects was shallow. With particular reference to forestry effects, Hartman *et al.* (1994) noted the need to recognize the high complexity of natural ecosystems, the interactions and differing timescales of relevant physical processes and the fact that the effects may differ between different species and life stages of fish. They added 'We believe that too much of it is done with little or no use of science'. These comments are valid within a much broader context. Many of the projects in England and Wales were based on subjective assessments.

There are also dangers of the science base being too narrow. The geomorphology, hydraulics and biology of the system need to be taken into account, together with the interests of all river (and catchment) users.

6.6.2   Focus

Mann & Winfield (1992) found that many projects were designed to give a general increase in habitat diversity rather than to rectify a specific defect. Even when specific defects were identified, these were often assumed to be detrimental, even though other possibilities had not been properly explored.

6.6.3 *Monitoring*

Hartman *et al.* (1994) commented that many projects suffered from weakness in the planning and development phase and from weakness or absence of experimental design to facilitate evaluation of the results. Mann & Winfield (1992) observed that only a minority of projects had the benefit of fish population surveys before and after restoration and/or some other measure of the effectiveness of the restoration (Table 6.5). A proper monitoring programme is an essential element of any restoration or improvement project and it should contain pre-intervention and post-intervention studies of fish populations by standardized, objective methods. It is also essential for both pre- and post-intervention studies to be of sufficient duration as to take account of year-upon-year variations (e.g. Figs 6.1 and 6.2) and for the post-intervention study to take account of any delayed effects. Hartman *et al.* quote Hunt (1976) to the effect that projects targeted on salmonids may take six or seven years before changes in population become fully apparent. Failure to appreciate the importance of year-upon-year variations is also mentioned by Mann & Winfield. The pre- and post-intervention studies need to be planned very carefully from the outset, to ensure that the data are of adequate quality and that the design of the work is sufficiently robust as to facilitate the types of statistical analysis needed to test, objectively, the effectiveness (or otherwise) of the treatment. This type of pre- and post-intervention study takes time and is relatively expensive. It therefore consumes funds that might otherwise contribute to the restoration work. This is probably the main reason for the inadequate monitoring observed in most projects but it is arguable that if the price of proper monitoring is a reduction in the number of projects started, then in the long run that price is worth paying. Proper pre-intervention investigations of high scientific standard are essential in planning the restoration/improvement in a focused manner and sound assessment of the

Table 6.5   Summary of deficiencies in documentation and monitoring of restoration projects in England and Wales. Based on Mann & Winfield (1992).

Deficiency	Species/group	Percentage frequency
No specific survey of fish populations before restoration	Salmon Trout	76.5 70.4
No specific survey of fish populations after restoration	Salmonids	68.2
No measure of effect of restoration	Salmon Trout	64.7 55.6

outcome of the work is essential in order to demonstrate its effectiveness. Without clear focus and adequate montoring, restoration is based on subjective judgement and is not capable of objective evaluation.

6.6.4   Recording and dissemination

Hartman *et al.* (1994) note that a record of project progress must be made public and that this requires sound evaluation of achievements. Mann & Winfield (1992) drew particular attention to the lack in many projects of costings, cost-benefit analyses or information on the costs and frequency of maintenance of structures. They emphasised particularly the need for full documentation of all aspects of each project. The necessary documentation can be outlined under ten headings:

(1) A detailed, objective (if possible, quantitative) assessment of the status of the fishery. This should ideally be on a catchment basis but should certainly be within a catchment context.

(2) An identification and, if possible, quantification, of habitat deficiencies for each life stage of each relevant salmonid species.

(3) A clear statement of objectives in terms of those aspects of habitat to be improved, in the light of (2) above.

(4) A clear and detailed plan of action, including costings and details of methodology.

(5) A detailed statement of the philosophy, methodology and results of a pre-intervention habitat survey.

(6) A detailed statement of the philosophy, methodology and results of a pre-intervention survey of the target fish species and life stages. This survey must be of adequate duration, precision and scientific rigour to be used in later assessment of the effectiveness of the project. At this point the statistical tests to be used in this pre- and post-intervention comparison must be clearly stated, preferably after consultation with a competent statistician.

(7) Full details of the execution of the work, including timing, costs, technical details, problems and variations of the original plan.

(8) A detailed statement of the philosophy, methodology and results of a post-intervention survey of the target fish species and life stages. This should be of sufficient duration that it can be used in later assessment of the effectiveness of the work. This study should be compatible with the one outlined in (6) above.

(9) A comparison of the pre- and post-intervention fish surveys ((6) and (8) above) and of the original objectives ((3) above) to assess the biological effectiveness of the work.

(10)   Use of the assessment of effectiveness ((9) above) and costs ((7) above) in cost-benefit analysis.

Lack of this sort of documentation inhibits the communication of experience and the process of learning. This is true even for failed projects because much can be learned from past mistakes, provided that they are properly documented. Reinvention of the wheel can be costly, especially when a proportion of the prototypes turns out to be square! The point was emphasised by Kondolf (1998) who quoted information to the effect that evaluation of habitat enhancement schemes in western North America (mainly centred upon salmonid habitat) indicated a 50% failure rate.

All the documentation in the world is of little value unless it is easily available and carries some hallmark of quality. The criterion of availability can be met by bringing together, in a hard copy archive, the documentation listed above and the raw data from the relevant surveys. Copies of the archive can then be deposited in places of good public access (a public library, or the library of a research institute) and the fact can be publicized. This should, perhaps, be the minimum requirement for any project whose proponents wish it to be taken seriously. It certainly goes some way to meeting the right of those providing public or private funding to see what has been done with their money. The requirement for quality assurance can be met if, in addition to the archive, a briefer and scientifically rigorous account of the project is published in a mainstream scientific journal. The advantages of this are threefold:

(1)   Before acceptance for publication in these journals, papers are reviewed by the editor and by one or more independent referees for presentation and scientific quality. This goes a long way towards ensuring that if accepted for publication, the work has a sound scientific basis and the conclusions are drawn objectively.
(2)   The published material is readily available to all interested parties via purchase of the journal, consultation of a library copy, or request of a reprint or photocopy from the authors.
(3)   The names and addresses of the authors are published with the paper and this facilitates the follow-up, by correspondence, of any points arising.

In recent years a multidisciplinary international programme has attempted physical restoration work on reaches of three lowland rivers, two in England and one in Denmark. This work was not related to salmonid habitat. Nevertheless, publication in the primary literature of the technical details and early results (Holmes & Nielsen, 1998; Vivash *et al.*, 1998; Kronvang *et al.*, 1998; Hoffman *et al.*, 1998; Biggs *et al.*, 1998) exemplifies a number of ideals outlined in this chapter with regard to the planning, management, execution and monitoring of restoration/rehabilitation work and to the communication of the results.

Chapter 7
What Next?
(beck to basics)

This small chapter has no summary because it is, itself, a sort of summary. Several issues already raised are highlighted and expanded slightly and the author's personal views on priority needs for the future are indicated.

Whilst writing Chapter 6, the author began to ask himself whether or not he believed in habitat restoration. The answer, with minor changes in the wording but not the sense, is a quotation from Hartman *et al.* (1994): 'Much of this paper [chapter] has focused on problems of doing stream restoration and habitat improvement. We [I] do not wish to discourage this work'. Chapter 4 is the heart of the present book and its intention is to bring together information that can be used to quantify, as far as possible, the environmental and habitat requirements of trout and salmon. This is essential to the identification of habitat problems and to the formulation of solutions that are as firmly based in good science as the present state of knowledge allows. In Chapter 6 a more critical tone is evident. This is not intended to be dismissive, but rather to be an encouragement towards projects that are more soundly based in science, more convincingly presented and more accountable both to the public and to the scientific community.

7.1 Expanding and deepening the science base

7.1.1 Long-term population studies

Careful, consistent, long-term studies on populations of trout and salmon have formed the basis of the stock-recruitment curves that we already have, and have provided many of our insights into the population dynamics and the practical management of these fish as well as contributing fundamental ecological knowledge. Despite this there are still many unknowns in this area of science and there is, therefore, a need for continuation of existing studies of this type and for the fostering of new ones. Elliott (1994) presented several powerful arguments in favour of long-term population studies. A major problem is that in recent decades, the continuity of funding for long-term fundamental research has become more problematical. This has reduced the incentive and the opportunity for scientists to initiate long-term projects.

7.1.2   *Environmental requirements*

This issue was covered in some detail in Chapter 4 but it is apparent that there are still considerable gaps in our knowledge, which need to be filled if we are to have a sound basis for assessing and improving trout and salmon habitat. Two aspects appear to be particularly important and both occur at the interfaces between sciences or between different fields of biology.

First, there is difficulty quantifying the relationships between various spawning/incubation habitat variables and the survival of eggs and alevins. We understand many of the relevant variables and the manner in which they interact but the various mechanisms and interactions cannot yet be quantified properly. There is therefore a need for well-thought-out observational and experimental studies. These need to be co-operative studies between scientists from a number of disciplines to include expertise in chemistry, physics, hydraulics, sedimentology, physiology and ecology.

Second, Chapter 4 deals mainly with chemical and physical aspects of the habitat requirements of trout and salmon. It is clear that some of the contradictions and deficiences in our understanding arise from the manner in which these requirements are modified via the physiology and behaviour of the fish. In recent years much fruitful work on the interfaces of physiology, behaviour and ecology (especially in relation to the feeding, growth and smolting of salmonids) has been published by workers in the University of Glasgow (e.g. Huntingford *et al.*, 1988; Metcalfe *et al.*, 1989; Metcalfe & Thorpe, 1992; Huntingford & De Leaniz, 1997). Work of this type is likely to make further valuable contributions in the future.

7.1.3   *Quantification of impacts*

It is comparatively easy to list the types of ecological impacts that can be expected from a given type of human activity. It is often more difficult to quantify those impacts, even for a few carefully chosen examples. Published work, much of it from long-term studies, enables us to put some approximate values upon the impacts of some forestry and civil engineering activities (e.g. Robinson, 1980; Kirby *et al.*, 1991; Neal *et al.*, 1992; Royal Commission on Environmental Pollution, 1996). Agriculture and fish farming impacts are more difficult to quantify because of the diffuse and spatially variable nature of agricultural activity and the comparatively recent growth of intensive fish farming. By casual observation and by inference from existing information and data, we can be reasonably sure that the effects of agriculture on the aquatic environment are varied and widespread and can be substantial. For example, a recent report by the Scottish Environment

Protection Agency (1999) stated that at present the main threat to Scotland's water environment is from sewage discharges, but that if present trends continue, agricultural run-off will be the biggest threat by the year 2010. There is a real need for better quantification.

7.2   Towards better management

In 1995 the chemical quality of 91% of rivers in England and Wales was 'very good to fair', 8% were 'poor' and 1% 'bad'. This represents an upgrading since 1990 of about 40% of total river length and a down-grading of about 12%, hence a nett improvement of about 28%. The corresponding nett improvement in biological quality was 26% (Environment Agency, 1998a). This is a striking illustration of the manner in which the water quality of British rivers has improved in recent years. It partly reflects better management and partly the decline of heavy industry. We must, however, interpret these figures with some caution. They probably indicate a very welcome reduction in continuous or regular discharges from point sources. It is likely, however, that in recent times there has been an increase of sporadic inputs, and of inputs from diffuse sources.

Discharges over brief time periods are unlikely to be detected in routine spot samples taken at infrequent (weekly or monthly) intervals and the sources of seepages from landfill sites or of downwash from agricultural land can be difficult to detect, locate and control.

Similarly, we have already noted the harmful effects of high concentrations of inert suspended solids on river biota, including salmonids. These solids are difficult to quantify or control and it is doubtful whether or not present legislation fully accepts such materials as 'pollutants'.

There is a need for a pro-active approach to sporadic and/or non-point-source pollution and to the problem of suspended solids. This implies a holistic view of land and water use and the need for an intensive programme of education, information and advice.

The recent publication (Raven *et al.*, 1998) of a summary of river habitat quality in the UK and Isle of Man develops a methodology and sets a baseline for future comparisons of habitat quality. The analysis takes into account a wide range of features of the general river habitat and is likely to be a valuable assessment tool in the future.

7.3   Improving monitoring, documentation and accountability

At present the improvement/restoration of rivers and of fish habitat is a growth industry and substantial sums of public and private money are going into a variety of projects. There is an obvious and understandable desire to spend as much as possible of the budget on the improvement work. This often leaves very little funding for proper assessment of the outcomes and for cost/benefit analysis. This is unfortunate because it is important to ensure that those funding the work get good value for their investment and that this is demonstrated, objectively, for each project. Unless these aims can be attained, restoration projects may lose their credibilty and may be perceived as bottomless pits into which money is thrown in the hope of useful results but without any objective proof that such results have been achieved. It is important to recognize that although science gives us some indications of suitable procedures, we do not always know how best to go about river restoration/improvement. We also do not know which approach will give the best return for money. Each project is therefore a learning experience and we should seek to maximize the amount that we learn from it. General restoration works on the Rivers Skerne and Cole (UK) began in 1994 and had a total budget of nearly £1 million. By 1998, 38% of this had been spent on physical works in rivers and floodplains; 20% on survey, design, documentation and supervision; 20% on management, promotion, compensation and administration; and 22% on 'before and after' monitoring (Holmes & Nielsen, 1998). We have already noted (Chapter 6) the long timescales needed for adequate biological monitoring; Biggs *et al.* (1998) observed, after a three-year post-restoration study of macroinvertebrates in the Rivers Skerne and Cole, that 'a longer period of study is required to determine the ultimate effects ... on the biota...'. It is therefore likely that the ultimate cost of monitoring the outcome of the work effectively will exceed the cost of the work itself. This is not an argument against proper monitoring but rather an indication that it is a costly but essential part of the process. There is a need for a proper balance between the costs of restoration and the costs of monitoring. Hitherto, there has been a gross imbalance towards the former in most projects in Europe and North America.

No amount of expensive monitoring is of much value unless the raw data are recorded and archived, and the results analysed adequately. Even this is of limited value unless hard copy of the raw data is deposited in a place to which the scientific community has ready access and the results are published in a quality journal that is readily accessible. The process of dissemination of knowledge in the UK has been hindered in recent years by a huge growth in the quantity of scientific and quasi-scientific material that is published in the 'grey literature'. This consists of a large number of internal reports, booklets and other documentation that has not been peer-reviewed and is not readily available to the scientific community for one or more of three reasons. First, only limited circulation may be authorized. Second, the scientific community may have difficulty learning of its existence. Third, the publication may be on sale at a comparatively high price which is an add-on cost to the scientist or to the library of his institute, over and above the on-going costs of mainstream scientific journals. It would be a major aid to scientific progress if this trend towards 'grey publication' could be reversed.

Appendix A
Mean instantaneous rate of mortality or loss

In a closed system, a decrease in numbers over time of a given cohort of fish can be taken as 'mortality'. In an open system it is better to consider it as 'loss', such that

$$\text{loss} = \text{mortality} + \text{emigration} - \text{immigration} \qquad (24)$$

though the general principles are still the same. A plot of survivors against time is usually a curve (Fig. A.1(a)). Very often, however, a given cohort of fish will approximate to a relatively constant rate of instantaneous mortality over a period of months or, even, years. In such cohorts a plot of ln (number of survivors) against time will approximate a straight line (Fig. A.1(b)) and the gradient of that

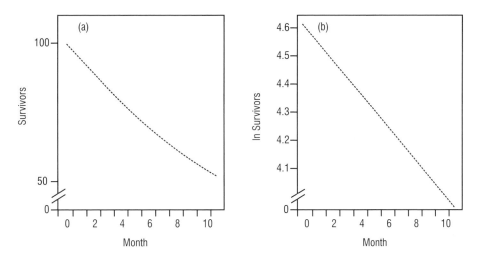

Fig. A.1 Survivorship graphs for a hypothetical cohort of fish. (a) A plot on natural scales, giving a concave curve that slopes downward from left to right. (b) A plot of the same data as (a) but using the natural logarithm (ln or logarithm to the base 'e') of the number of survivors. This gives a straight line of negative gradient. The equation of this line is ln (survivors) = $-$ m (month number) + c, where m (gradient of the line with sign changed) is the mean instantaneous mortality rate per month and c is the estimated value of ln (survivors) in month 0.

line, with sign changed, is an estimate of the mean rate of instantaneous mortality per unit time. In the example of Fig. A.1, the timescale is in months and the instantaneous rate of mortality derived from the gradient of the fitted line is therefore the rate per month. In the example shown, the mean instantaneous rate per month is 0.0594.

These instantaneous rates have two valuable properties. First, the mean rate per year (assuming the rate does not show appreciable seasonal variations) will be 12 times the rate per month and the daily rate will be $\frac{1}{30}$ (or $\frac{1}{31}$) of the monthly rate. Second, components of the total rate of mortality (Z) are additive. Hence, if F is the instantaneous rate of fishing mortality and M is the instantaneous rate of natural mortality, then

$$Z = M + F \tag{25}$$

The use of the word 'instantaneous' may be a little confusing for some readers. Strictly, these rates should be computed and applied over very short periods of time (hours or days). In practice, their convenience and utility is such that they are frequently applied over longer periods of time and referred to as 'mean' rates.

Appendix B
Mean instantaneous rate of growth (in weight)

The mean instantaneous growth rate uses the natural logarithm (logarithm to the base 'e'), ln, and is really a representation of growth as a process akin to compound interest. If a theoretical fish starts at a weight of z grams and adds 5% to its weight each day, then after one day its weight will be

$$z + 0.05\ z = 1.05\ z \tag{26}$$

grams. After two days its weight will be

$$1.05\ z + (0.05 \times 1.05\ z) = 1.1025\ z \tag{27}$$

grams, rather than the

$$1.0\ z + 0.05\ z + 0.05\ z = 1.10\ z \tag{28}$$

grams that we would expect from simple interest.

If, on day 0, a fish weighs w_0 grams and after t days it weighs w_t grams, then the mean instantaneous rate of growth in weight (G) is given by the equation

$$G\ day^{-1} = (\ln w_t - \ln w_0)/t \tag{29}$$

The weight of the fish after t days at mean instantaneous growth rate G would be given by

$$\ln w_t = \ln w_0 + t\ G \tag{30}$$

As with instantaneous rate of mortality, mean weekly or monthly rates can be estimated from daily rates by multiplication of the daily rate by seven or thirty (or thirty one), respectively.

Glossary

adipose fin	A single small fatty fin situated on the back of a salmonid posterior to the dorsal fin (*see* Fig 1.1).
afon	A river or large stream (Wales).
ages of salmon and trout	To attribute ages to a given population of salmon or trout in fresh water it is necessary, first, to attribute a nominal birthday. In northern Britain spawning is usually in the autumn or early winter and swim-up is usually in the spring of the following year. The birthday is then taken as, for example, 1 April or 1 May in the year of swim-up. At that point the fish is assumed to be 0 years of age. Between then and the birthday's first anniversary the age is given as '0+' or 'age group 0' or '0-group'. In the following year the age is '1+' or 'age group I' or 'I-group' and so on through subsequent years. In this book the 'I-group' convention has been followed. When salmonids go to sea it is usual to describe their age in terms of the number of winters spent in fresh water and in the sea. A fish of 2.3 winters would have spent two winters in fresh water, smolted as a II-group fish and then spent three winters at sea, e.g. *see* Table 2.3. The fish of a given population that swam up in the same year as one another are referred to as a 'cohort' or 'year class'.
alevin	Embryo of a trout or salmon after hatching from the egg and before emergence from the gravel (*see* Chapter 2, p. 16).
allochthonous	An adjective to describe material that enters a stream or river from the terrestrial environment.
anadromous	An adjective to describe fish that live the early part of their lives in fresh water, go to sea and then return to fresh water to spawn.
anal fin	A single fin on the ventral side of the fish, posterior to the anus. *See* Fig 1.1.

anoxia	Lack of oxygen.
base flow	That part of the flow in a stream or river that is contributed by ground water.
Base flow index	An index of the relative contribution of base flow to the total flow in a stream. (*See* Chapter 3, p. 39).
beck	A stream (northern England).
burn	A stream (Scotland and northern England).
carrying capacity	The maximum number or population density of fish of a given species at a given age or size that can be supported by a given area or portion of a stream or river. The value may vary appreciably from one year to another and may be modified by the numbers of fish of other species or of other life stages of the same species.
catchment	The tract of land from which water, falling as rain or snow, drains into a watercourse. The perimeter of the catchment is described as a 'watershed'. (In north America, 'watershed' is used to describe a catchment.)
Chironomidae	Non-biting midges.
cohort	*See* 'ages of salmon and trout'.
critical period	In salmonid populations that have produced large numbers of fry, the period soon after territories are occupied is the 'critical period' in which numbers are reduced by density-dependent mortality/dispersal to match the carrying capacity of the habitat.
discharge	1. Release of water from a reservoir, or of effluent into a watercourse. 2. Volume of water passing a point in a watercourse per unit of time.
dorsal fin	The main fin on the back of a salmonid. *See* Fig 1.1.
Ephemeroptera	Mayflies.
epilimnion	*See* 'stratification of lakes'.
fines	Fine material in a bed of a stream or river. The precise size specified for 'fines' varies greatly between authors. (*See* Chapter 3, p. 57–58).
fry	A young salmonid after emergence from the gravel and before assumption of territory. (*See* Chapter, 2 p. 16).
grilse	A salmon that has spent one winter at sea and then returned to fresh water as a mature fish.
Habscore	A mathematical model used to assess salmonid habitat quality and utilization on the basis of a set of

field measurements of habitat. (*See* Chapter 6, pp. 149, 150).

hypoxia
Low oxygen concentration.

instantaneous rates
Logarithmic rates of growth or mortality. *See* Appendices A and B.

kelt
An adult salmon or sea trout after spawning.

lateral line
A line of special scales running along the flank of the fish. These scales are associated with sensory organs. (Shown, but not labelled, on Fig 1.1).

ln
See 'logarithm'.

$\log_{10}$
See 'logarithm'.

logarithm
For a detailed account a mathematics text should be consulted. Three types of logarithm are mentioned in this book:

(1) 'Common logarithms' or logarithms to the base 10; usually indicated by '$\log_{10}$'.
(2) 'Natural logarithms,' also known as Naperian logarithms or logarithms to the base 'e' (e = 2.7183). Usually indicated by 'ln' and used in the calculation of instantaneous rates of mortality and growth (*see* Appendices A and B).
(3) The 'phi scale' is a scale of logarithms to the base 2 and is used to describe the composition of gravels (*see* Chapter 3, p. 56).

microcrustacea
Small Crustacea, especially water fleas (Cladocera), cyclops (Copepoda) and cypris (Ostracoda).

ninety-five percent confidence limits
Usually written as '95% C.L'. These limits are attached to an estimated value and indicate that there is a 95% probability that the true value lies within the 95% C.L. For example, if the value of a variable is estimated as 100 units and the 95% C.L are $\pm$ 10 units, then there is a 95% probability that the true value of the variable is between 100 + 10 = 110 and 100 − 10 = 90 units.

parr
Young salmonid after assumption of a territory and before smolting(*see* Chapter 2, p. 16).

pectoral fins
Paired fins inserted on the side of the fish just posterior to the operculum in salmonids (*see* Fig 1.1).

percentile
Best defined by use of an example. In Table 3.4, the concentration limits for nickel in soft water are given as an average of $< 10 \ \mu g\,l^{-1}$ with a 95 percentile of

	$<30\,\mu g\,l^{-1}$. This implies the average concentration must be less than $10\,\mu g\,l^{-1}$ and that the concentration should be less than $30\,\mu g\,l^{-1}$ for at least 95% of the time.
pH	A logarithmic scale to indicate the acidity or alkalinity of water (*see* Chapter 3, p. 42).
PHABSIM	'Physical Habitat Simulation'. A mathematical model that is used to assess fluvial habitat quality on the basis of water depth and velocity, substratum and cover (*see* Chapter 6, pp. 148, 149).
'phi' scale	*See* 'logarithm'.
porosity	An index of the porousness of a gravel deposit. Denoted by the symbol λ and calculated from the fraction of a unit of cross-section of gravel that consists of void (*see* Chapter 3, p. 59 and equation (6)).
redd	The gravel nest in which a female salmonid buries her eggs.
resident	Used to describe salmonids that remain all their lives in a stream or river system rather than migrating downstream to mature in a lake or the sea.
sike	A stream (northern England).
smolt	A juvenile salmon or sea trout that is ready to migrate to the sea and has acquired the characteristic appearance of a smolt as described in Chapter 2, pp. 16, 21.
stratification of lakes	The floating of a layer of warm, well-oxygenated water (the 'epilimnion') upon a layer of cool, often poorly oxygenated water (the 'hypolimnion'). This occurs in some lakes during summer. The thin transition zone between epilimnion and hypolimnion is described as the 'thermocline' (*see* Chapter 3, p. 51).
swim-up	The emergence of a salmonid fry from the gravel to become a free-swimming fish feeding entirely on external foods (*see* Chapter 2, p. 16).
thermocline	*See* 'stratification of lakes'.
year class	*See* 'ages of salmon and trout'.

References

Alabaster, J.S. (1970) River flow and upstream movement and catch of migratory salmonids. *Journal of Fish Biology*, **2**, 1–13.

Alabaster, J.S. (1972) Suspended solids and fisheries. *Proceedings of the Royal Society, London B*, **180**, 395–406.

Alabaster, J.S. (1986) An analysis of angling returns for trout, *Salmo trutta* L., in a Scottish river. *Aquaculture and Fisheries Management*, **17**, 313–16.

Alabaster, J.S. (1990) The temperature requirements of adult salmon, *Salmo salar* L., during their upstream migration in the River Dee. *Journal of Fish Biology*, **37**, 659–61.

Alabaster, J.S. & Durbin, F.J. (1965) *Blood groups in salmon, trout and their hybrids.* Report, pp. 38–9. Salmon Research Trust of Ireland 1964.

Alabaster, J.S., Gough, P.J. & Brooker, W.J. (1991) The environmental requirements of salmon, *Salmo salar* L., during their passage through the Thames Estuary, 1982–1989. *Journal of Fish Biology*, **38**, 741–62.

Alabaster, J.S. & Lloyd, R. (eds) (1982) *Water Quality Criteria for Freshwater Fish*. Butterworths, London.

Allan, I.R.H. & Ritter, J.A. (1977) Salmon terminology. *Journal du Conseil International pour l'Exploration de la Mer*, **37**, 293–9.

Allen, K.R. (1940) Studies on the biology of the early stages of salmon (*Salmo salar*). I. Growth in the River Eden. *Journal of Animal Ecology*, **9**, 1–23.

Allen, K.R. (1944) Studies on the biology of the early stages of the salmon (*Salmo salar* L.). 4. The smolt migration in the Thurso River in 1938. *Journal of Animal Ecology*, **13**, 63–85.

Allen, K.R. (1985) Comparison of the growth rate of brown trout, *Salmo trutta*, in a New Zealand stream with experimental fish in Britain. *Journal of Animal Ecology*, **54**, 487–95.

Anon. (1992) *Restoration of Aquatic Ecosystems*. Science, Technology and Public Policy. National Research Council, Washington, D.C., Water Science and Technology Board.

Aprahamian, M.W., Jones, G.O. & Gough, P.J. (1998) Movement of Atlantic salmon in the Usk estuary, Wales. *Journal of Fish Biology*, **53**, 221–5.

Arrowsmith, E. & Pentelow, F.T.K. (1965) The introduction of salmon and trout to the Falkland Islands. *Salmon and Trout Magazine*, **174**, 119–29.

Backiel, T. & Le Cren, E.D. (1978) Some density relationships for fish population parameters. In: *Ecology of Freshwater Fish Production*, (ed. Gerking, S.D.) pp. 279–302. Blackwell Scientific Publications, Oxford.

Bagenal. T.B. (1969) The relationship between food supply and fecundity in brown trout *Salmo trutta* L. *Journal of Fish Biology*, **1**, 167–82.

Baglinière J–L. (1976) Étude des populations de saumon atlantique (*Salmo salar* L.) en Bretagne–Basse Normandie. II. Activité de devalaison des smolts sur l'Elle. *Annales Hydrobiologie*, **7**, 159–77.

Baglinière, J–L. & Arribe–Moutounet, D. (1985) Microréparation des populations de truite commune (*Salmo trutta* L.) de juvéniles de saumon atlantique (*Salmo salar* L.) et des autres espèces présentes dans la partie haute du Scorff (Bretagne). *Hydrobiologia*, **120**, 229–89.

Baglinière, J-L. & Maisse, G. (eds) (1991) *La truite biologie et écologie*. INRA, Paris.

Bams, R.A. (1969) Adaptations of sockeye salmon associated with incubation in stream gravel beds. In: *Symposium on salmon and trout in streams* (ed. T.G. Northcote). H R MacMillan Lectures in Fisheries, Vancouver, 71–81.

Bandt, H.J. (1936) Der für Fische 'tödlich pH–wert' in alkelinischen Bereich. *Zeitschrift für Fischerei und deven Hilfswissenschafen*, **34**, 359–61.

Banks, J.W. (1969) A review of literature on the upstream migration of adult salmonids. *Journal of Fish Biology*, **1**, 85–136.

Baxter, G.C. (1961) River utilisation and the preservation of migratory fish life. *Proceedings of the Institute of Civil Engineers*, **18**, 225–44.

Beach, M.H. (1984) Fish passage design – criteria for the design and approval of fish passes and other structures to facilitate the passage of migratory fish in rivers. *M.A.F.F. Fisheries Research Technical Report*, **78**, 1–46.

Beall, E., Dumas, J., Claireaux, D., Barrière L. & Marty, C. (1994) Dispersal patterns and survival of Atlantic salmon (*Salmo salar* L.) juveniles in a nursery stream. *ICES Journal of Marine Science*, **51**, 1–9.

Beamish, F.H.W. (1978) Swimming capacity. In: *Fish Physiology Volume VII Locomotion* (eds W.S. Hoar & D.J. Randall). Academic Press, London 101–87.

Beiningen, K.T. & Ebel, W.J. (1970) Effect of John Day Dam on discharged nitrogen concentrations and salmon in the Columbia River. *Transactions of the American Fisheries Society*, **99**, 664–71.

Beland, T.G., Jordan, R.M. & Meister, A.L. (1982) Water depth and velocity preferences of spawning Atlantic salmon in Maine rivers. *North American Journal of Fisheries Management*, **2**, 11–13.

Bell, M.C. (1986) *Fisheries handbook of engineering requirements and biological criteria*. U.S. Army Corps of Engineers, Office of Engineers, Fish Passage and Development Program, Portland, Oregon.

Berg, O.K. & Berg, M. (1987) Migrations of sea trout, *Salmo trutta* L., from the Vardnes River in northern Norway. *Journal of Fish Biology*, **31**, 113–21.

Berg, O.K. & Moen, V. (1999) Inter- and intrapopulation variation in temperature sum requirements at hatching for Norwegian Atlantic salmon. *Journal of Fish Biology*, **54**, 636–47.

van den Berghe, E.P. & Gross, M.R. (1984) Female size and nest depth in coho salmon (*Oncorhynchus kisutch*). *Canadian Journal of Fisheries and Aquatic Sciences*, **41**, 204–6.

Beverton, R.J.H. & Holt, S.J. (1957) On the dynamics of exploited fish populations. *Fishery Investigations, Series 2*, **19**, 1–533. Ministry of Agriculture and Fisheries, London.

Bieniarz, K. (1973) Effect of light and darkness on incubation of eggs, length, weight and sexual maturity of sea trout (*Salmo trutta* L.), brown trout (*Salmo trutta fario* L.) and rainbow trout (*Salmo irideus*, Gibbons). *Aquaculture*, **2**, 199–315.

Biggs, J., Carfield, A., Grøn, P., Hansen, H.O., Walker, D., Whitfield, M. & Williams, P. (1998) Restoration of the Rivers Brede, Cole, and Skerne: Joint Danish and British EU–LIFE Demonstrations project V. Short–term impacts on the conservation value of aquatic macroinvertebrate and macrophyte assemblages. *Aquatic Conservation Marine and Freshwater Ecosystems*, **8**, 241–55.

Bilton, H.T., Alderdice, D.F. & Schnute, J.T. (1982) Influence of time and size at release of coho salmon (*Oncorhynchus kisutch*) on returns at maturity. *Canadian Journal of Fisheries and Aquatic Sciences*, **39**, 426–47.

Bisson, P.A., Reeves, G.G., Bilby, R.E. & Naiman, R.J. (1997) Watershed management and Pacific salmon: desired future conditions. In: *Pacific salmon and their ecosystems* (eds D.J. Stouder, P.A. Bisson & R.J. Naiman). Chapman & Hall, New York, 447–74.

Blair, W.F., Blair, A.P., Brodkorb, P., Cagle, F.R. & Moore, G.A. (1957) *Vertebrates of the United States*. McGraw-Hill, New York.

Blaxter, J.H.S. (1969) Swimming speeds of fish. *F.A.O. Fisheries Report*, **62**, 69–100.

Boeuf, G. (1986) La salmoniculture au Chili. *Pisciculture Francaise*, **84**, 5–35.

Bohlin, T. (1975) A note on aggressive behaviour of adult male sea trout towards precocious males during spawning. *Report, Institute of Freshwater Research, Drottningholm*, **54**, 118.

Bohlin, T. (1978) Temporal changes in the spatial distribution of juvenile sea trout *Salmo trutta* in a small stream. *Oikos*, **30**, 114–20.

Bohlin, T., Hanurin, S., Heggberget, T.G., Rasmussen, G. & Saltveit, S.J. (1989) Electrofishing – Theory and practice with special emphasis on salmonids. *Hydrobiologia*, **173**, 9–43.

Boon, P.J., Calow, P. & Petts, G.E. (1992) *River Conservation and Management*. John Wiley & Sons, Chichester.

Bovee, K.D. (1982) A guide to stream habitat analyses using Instream Flow Incremental Methodology. *FWS/OBS – 82/26 Office of Biological Sciences US Fish and Wildlife Service. Instream Flow Information Paper*, **12**.

Brännäs, E. (1988) Emergence of Baltic salmon, *Salmo salar* L., in relation to temperature: a laboratory study. *Journal of Fish Biology*, **33**, 589–600.

Brett, J.R. (1952) Temperature tolerance in young Pacific salmon, genus *Oncorhynchus*. *Journal of the Fisheries Research Board of Canada*, **9**, 265–323.

Brett, J.R. (1956) Some principles in the thermal requirements of fishes. *Quarterly Review of Biology*, **31**, 75–87.

Brookes, A. & Shields, F.D. (1996**a**) *River Channel Restoration. Guiding Principles for Sustainable Projects*. John Wiley & Sons, Chichester.

Brookes, A. & Shields, F.D. (1996**b**) Perspectives on river channel restoration. In: *River Channel Restoration Guiding Principles for Sustainable Projects* (eds A. Brookes & F.D. Shields). John Wiley & Sons, Chichester, New York, Brisbane, Toronto & Singapore.

Brown, D.J.A. (1983) Effect of calcium and aluminium concentrations on the survival of brown trout (*Salmo trutta*) at low pH. *Bulletin of Environmental Contamination and Toxicology*, **30**, 382–7.

Brown, D.J.A. & Lynam, S. (1981) The effect of sodium and calcium concentrations on the survival of brown trout (*Salmo trutta*) at low pH. *Bulletin of Environmental Contamination and Toxicology*, **30**, 582–7.

Buck, R.J.G. & Hay, D.W. (1984) The relation between stock size and progeny of Atlantic salmon *Salmo salar* L. in a Scottish stream. *Journal of Fish Biology*, **24**, 1–11.

Burger, C.V., Wilmot, R.L. & Wangaard, D.B. (1985) Comparison of spawning areas and times for two runs of chinook salmon (*Oncorhynchus tshawytscha*) in the Kenai River, Alaska. *Canadian Journal of Fisheries and Aquatic Sciences*, **42**, 693–700.

Butcher, R.W., Longwell, J. & Pentelow, F.J.K. (1937) Survey of the River Tees Part III – The Non–Tidal Reaches – Chemical and Biological. *Department of Scientific and Industrial Research Technical Paper*, **6**, 1–189.

Bye, V.J. (1984) The role of environmental factors in the timing of reproductive cycles. In: *Fish Reproductive Strategies and Tactics* (eds R.W. Potts & R.J. Wootton). Academic Press, London. 187–205.

Cairns, J. (1990) Lack of theoretical basis for predicting rate and pathways of recovery. *Environmental Management*, **14**, 517–26.

Cairns, J. (1991) The status of the theoretical and applied science of restoration ecology. *The Environmental Professional*, **13**, 186–94.

Capra, H., Breil, P. & Suchon, Y. (1995) A new tool to interpret magnitude and duration of fish habitat variations. *Regulated Rivers: Research and Management*, **10**, 281–9.

Carling, P.A. (1983) Threshold of coarse gravel sediment transport in broad and narrow natural streams. *Earth Surface Processes and Landforms*, **8**, 1–18.

Carling, P.A. & Dobson, J.H. (1992) Fish pass design and evaluation. Phase 1. Initial Review. *National Rivers Authority R & D Note*, **110**, 1–186.

Carling, P.A. & McCahon, C.P. (1987) Natural siltation of brown trout (*Salmo trutta* L.) spawning gravels during low flow conditions. In: *Regulated Streams: advances in ecology* (eds J.F. Craig & B.J. Kemper). Plenum, New York, 229–43.

Carling, P.A. & Reader, N.A. (1981) A freeze–sampling technique suitable for coarse river bed-material. *Sedimentary Geology*, **29**, 233–9.

Carling, P.A. & Reader, N.A. (1982) Structure, composition and bulk properties of upland stream gravels. *Earth Surface Processes and Landforms*, **7**, 349–65.

Carrick, T.R. (1979) The effect of acid water on the hatching of salmonid eggs. *Journal of Fish Biology*, **14**, 165–72.

Casey, H. (1969) The chemical composition of some southern English chalk streams and its relation to discharge. *Association of River Authorities Year Book 1969* 100–13.

Casey, H. & Newton, P.V. (1972) The chemical composition and flow of the South Winterbourne in Dorset. *Freshwater Biology*, **2**, 229–34.

Castleberry, D.T., Cech, J.J., Erman, D.C., Haskin, D., Healey, M., Kondolf, G.M., Mangel, M., Mohr, M., Moyle, P.B., Nielsen, J., Speed, T.P. & Williams, J.G. (1996) Uncertainty and instream flow standards. *Fisheries*, **21**, 20–21.

Central Fisheries Board (1995a) *Habitat improvement for juvenile salmon and trout in small streams.* Central Fisheries Board, Dublin.

Central Fisheries Board (1995b) *Restoration of some essential natural physical character-istics of salmon and trout streams.* Central Fisheries Board, Dublin.

Chadwick, E.M.P. (1985) The influence of spawning stock on production and yield of Atlantic salmon, *Salmo salar* L., in Canadian rivers. *Aquaculture and Fisheries Management*, **16**, 111–9.

Chapman, D.W. (1988) Critical review of variables used to define effects of fines in redds of large salmonids. *Transactions of the American Fisheries Society*, **117**, 1–21.

Clarke, W.C. & Hirano, T. (1995) Osmoregulation. In: *Physiological Ecology of Pacific Salmon* (eds C. Groot, L. Margolis & W.C. Clarke). UBC Press, Vancouver, 317–77.

Clay, C.H. (1961) *Design of fishways and other fish facilities*. Department of Fisheries, Canada, Ottawa.

Coble, D.W. (1961) Influence of water exchange and dissolved oxygen in redds on survival of steelhead trout embryos. *Transactions of the American Fisheries Society*, **90**, 469-74.

Cooper, A.C. (1965) The effects of transported stream sediments on the survival of sockeye and pink salmon eggs and alevins. *International Pacific Salmon Fisheries Commission Bulletin*, **18**, 1–71.

Cowx, I.G. (1994a) *Rehabilitation of Freshwater Fisheries*. Fishing News Books, Oxford.

Cowx, I.G. (1994b) Strategic approach to fishery rehabilitation. In: *Rehabilitation of Freshwater Fisheries* (ed. I.G. Cowx). Fishing News Books, Oxford, 3–10.

Cowx, I.G. (1995) Fish Stock Assessment – A biological basis for sound ecological management. In: *The Ecological Basis for River Management* (eds D.M. Harper & A.J.D. Ferguson). John Wiley & Sons, Chichester, 375–88.

Cowx, I.G. & Welcomme, R.L. (1998) *Rehabilitation of rivers for fish. A study undertaken by the European Inland Fisheries Advisory Commission of F.A.O.* Fishing News Books, Oxford.

Cragg–Hine, D. (1985) The assessment of flow requirements for upstream migration of salmonids in some rivers of North-West England. In: *Habitat Modification and Freshwater Fisheries* (ed. J.S. Alabaster). Butterworths, London, 209–15.

Crisp, D.T. (1966) Input and output of minerals for an area of Pennine moorland: the importance of precipitation, drainage, peat erosion and animals. *Journal of Applied Ecology*, **3**, 327–348.

Crisp, D.T. (1977) Some physical and chemical effects of Cow Green (Upper Teesdale) impoundment. *Freshwater Biology*, **7**, 109–20.

Crisp, D.T. (1981) A desk study of the relationship between temperature and hatching time for the eggs of five species of salmonid fishes. *Freshwater Biology*, **11**, 361–8.

Crisp, D.T. (1984) Effects of Cow Green Reservoir upon downstream fish populations. *Freshwater Biological Association Annual Report*, **52**, 47–62.

Crisp, D.T. (1985) Thermal 'resetting' of streams by reservoir releases with special reference to effects on salmonid fishes. In: *Regulated Streams: advances in ecology* (eds J.F. Craig & J.B. Kemper). Plenum, New York, 163–82.

Crisp, D.T. (1988a) Water temperature data for streams and rivers in northeast England. *Freshwater Biological Association Occasional Publication*, **26**, 1–60.

Crisp, D.T. (1988b) Prediction from water temperature, of eyeing, hatching and swim-up times for salmonid embryos. *Freshwater Biology*, **19**, 41–8.

Crisp, D.T. (1989) Use of artificial eggs in studies of washout depth and drift distance for salmonid eggs. *Hydrobiologia*, **178**, 155–63.

Crisp, D.T. (1990a) Some effects of application of mechanical shock at varying stages of development upon the survival and hatching time of British salmonid eggs. *Hydrobiologia*, **194**, 57–65.

Crisp, D.T. (1990b) Water temperature within a stream gravel bed and implications for salmonid incubation. *Freshwater Biology*, **23**, 601–12.

Crisp, D.T. (1991) Stream channel experiments on downstream movement of recently emerged trout, *Salmo trutta* L., and salmon, *S. salar* L. – III. Effects of developmental stage and day and night upon dispersal. *Journal of Fish Biology*, **39**, 371–81.

Crisp, D.T. (1992) Measurement of stream water temperature and biological applications to salmonid fishes, grayling and dace (including ready reckoners). *Freshwater Biological Occasional Publication*, **29**, 1–72.

Crisp, D.T. (1993a) Population densities of juvenile trout (*Salmo trutta*) in five upland streams and their effects upon growth, survival and dispersal. *Journal of Applied Ecology*, **30**, 759–71.

Crisp, D.T. (1993b) The environmental requirements of salmon and trout in fresh water. *Freshwater Forum*, **3**, 176–202.

Crisp, D.T. (1993c) The ability of UK salmonid alevins to emerge through a sand layer. *Journal of Fish Biology*, **43**, 656–8.

Crisp, D.T. (1994) Reproductive investment of female brown trout, *Salmo trutta* L., in a stream and reservoir system in northern England. *Journal of Fish Biology*, **44**, 343–9.

Crisp, D.T. (1995) Dispersal and growth rate of 0-group salmon (*Salmo salar* L.) from point-stocking together with some information from scatter stocking. *Ecology of Freshwater Fish*, **4**, 1–8.

Crisp, D.T. (1996a) Environmental requirements of common riverine European salmonid fish species in fresh water with particular reference to physical and chemical aspects. *Hydrobiologia*, **323**, 201–21.

Crisp, D.T. (1996b) Experimental studies on planting artificial stream channels with unfed and six weeks fed salmon (*Salmo salar* L.) and trout (*S. trutta* L.) fry/parr. *Ecology of Freshwater Fish*, **5**, 68–75.

Crisp, D.T. (1998) Water temperature of Plynlimon streams. *Hydrology and Earth System Sciences*, **1**, 535–40.

Crisp, D.T. & Beaumont, W.R.C. (1995) The trout (*Salmo trutta*) population of the Afon Cwm, a small tributary of the Afon Dyfi, mid-Wales. *Journal of Fish Biology*, **46**, 703–16.

Crisp, D.T. & Beaumont, W.R.C. (1996) The trout (*Salmo trutta* L.) populations of the headwaters of the Rivers Severn and Wye, mid-Wales, UK. *The Science of the Total Environment*, **177**, 113–23.

Crisp, D.T. & Beaumont, W.R.C. (1998) Fish populations in Plynlimon streams. *Hydrology and Earth System Sciences*, **1**, 541–8.

Crisp, D.T. & Carling, P.A. (1989) Observations on siting, dimensions and structure of salmonid redds. *Journal of Fish Biology*, **34**, 119–34.

Crisp, D.T. & Cubby, P.R. (1978) The populations of fish in tributaries of the River Eden on the Moor House National Nature Reserve, northern England. *Hydrobiologia*, **57**, 85–93.

Crisp, D.T. & Gledhill, T. (1970) A quantitative description of the recovery of the bottom fauna in a muddy reach of a mill stream in Southern England after draining and dredging. *Archiv für Hydrobiologie*, **67**, 502–41.

Crisp, D.T. & Howson, G. (1982) Effect of air temperature upon mean water temperature in streams in the north Pennines and English Lake District. *Freshwater Biology*, **12**, 359–67.

Crisp, D.T. & Hurley, M.A. (1991a) Stream channel experiments on downstream movement of recently emerged trout, *Salmo trutta* L., and salmon, *S. salar* L. – I. Effects of four different water velocity treatments upon dispersal rate. *Journal of Fish Biology*, **39**, 347–61.

Crisp, D.T. & Hurley, M.A. (1991b) Stream channel experiments on downstream movement of recently emerged trout, *Salmo trutta* L., and salmon, *S. salar* L. – II. Effects of constant and changing velocities and of day and night upon dispersal. *Journal of Fish Biology*, **39**, 363–70.

Crisp, D.T & Mann, R.H.K. (1977a) Analysis of fishing records from Cow Green Reservoir, upper Teesdale, 1971–1975. *Fisheries Management*, **8**, 23–34.

Crisp, D.T. & Mann, R.H.K. (1977b) A desk study of the performance of trout fisheries in a selection of British Reservoirs. *Fisheries Management*, **8**, 101–19.

Crisp, D.T., Mann, R.H.K., Cubby, P.R. & Robson, S. (1990) Effects of impoundment upon trout (*Salmo trutta* L.) in the basin of Cow Green Reservoir. *Journal of Applied Ecology*, **27**, 1020–41.

Crisp, D.T., Mann, R.H.K. & McCormack, Jean C. (1974) The populations of fish at Cow Green, upper Teesdale before impoundment. *Journal of Applied Ecology*, **11**, 969–96.

Crisp, D.T., Mann, R.H.K. & McCormack, Jean C. (1975) The populations of fish in the River Tees on the Moor House National Nature Reserve, Westmorland. *Journal of Fish Biology*, **7**, 573–93.

Crisp, D.T., Matthews, A.M. & Westlake, D.F. (1982) The temperatures of nine flowing waters in southern England. *Hydrobiologia*, **6**, 54–65.

Crisp, D.T. & Robson, S. (1979) Some effects of discharge upon the transport of animals and peat in a north Pennine headstream. *Journal of Applied Ecology*, **16**, 721–36.

Crisp, D.T & Robson, S. (1982) Analysis of fishing records for Cow Green Reservoir, upper Teesdale, 1971–1980. *Fisheries Management*, **13**, 65–78.

Cross, T.F. (1989) Genetics and Management of the Atlantic Salmon. *Atlantic Salmon Trust*, 1–73.

Crozier, W.W. (1998) Genetic implications of hatchery rearing in Atlantic salmon: effects of rearing environment on genetic composition. *Journal of Fish Biology*, **52**, 1014–25.

Crozier, W.W. & Kennedy, G.J.A. (1995) The relationship between a summer fry (0+) abundance index, derived from semi–quantitative electrofishing and egg deposition of Atlantic salmon in the River Bush, Northern Ireland. *Journal of Fish Biology*, **47**, 1055–62.

Davis, J.C. (1975) Minimal dissolved oxygen requirements of aquatic life with emphasis on Canadian species: a review. *Journal of the Fisheries Research Board of Canada*, **32**, 2295–332.

Davies, J.K. (1988) A review of information relating to fish passage through turbines: implications to tidal power schemes. *Journal of Fish Biology*, **33 (A)**, 111–26.

De Gaudemar, B. & Beall, E. (1998) Effects of overripening on spawning behaviour and reproductive success of Atlantic salmon females spawning in a controlled flow channel. *Journal of Fish Biology*, **53**, 434–66.

De Graaf, D.A. & Bain, L.H. (1986) Habitat use preferences of juvenile Atlantic salmon in two Newfoundland rivers. *Transactions of the American Fisheries Society*, **115**, 671–81.

De Vries, P. (1997) Riverine salmonid egg burial depths: review of published data and implications for scour studies. *Canadian Journal of Fisheries and Aquatic Sciences*, **54**, 1685–98.

Department of Agriculture and Fisheries for Scotland (1982a) *Report on Survey of Fodder Silos*. Report prepared by N. Taylor & T.G. Brownlie. Department of Agriculture and Fisheries for Scotland.

Department of Agriculture and Fisheries for Scotland (1982b) *A Study of the Economic value of Sporting Salmon Fishing in Three areas of Scotland*. Report prepared by Tourism and Recreation Unit, Edinburgh University.

Department of the Environment, Transport and the Regions (1998) *The review of the water abstraction licensing system in England and Wales*. Consultation Paper. Department of the Environment, Transport and the Regions, Welsh Office.

Department of the Marine (1993) *Report of the sea trout working group*. Fisheries Research Centre, Abbotstown, Eire.

Donaghy, M.J. & Verspoor, E. (1997) Egg survival and timing of hatch in two Scottish Atlantic salmon stocks. *Journal of Fish Biology*, **51**, 211–14.

Drakeford, T. (1981) *Management of upland streams. An experimental fisheries management project on the afforested headwaters of the River Fleet, Kirkudbrightshire*. Institute of Fisheries Management. Annual Study Course, Durham, 1981, 86–92.

Dumont, D. & Mongeau, J.R. (1989) La truite brune (*Salmo trutta*) dans le Québec méridional. *Bulletin Francaise de Pisciculture*, **319**, 153–66.

Dwyer, W.P, Frendenberg, W. & Erdahl, D.A. (1993) Influence of electroshock and mechanical shock on survival of trout eggs. *North American Journal of Fishery Management*, **13**, 839–43.

Edwards, R.W. & Crisp, D.T. (1982) Ecological implications of river regulation in the United Kingdom. In: *Gravel-bed Rivers* (eds R.D. Hey, J.C. Bathurst & C.R. Thorne). John Wiley & Sons, Chichester, 843–65.

Edwards, R.W., Densem, J.W. & Russell, P.A. (1979) An assessment of the importance of temperature as a factor controlling growth rate of trout in streams. *Journal of Animal Ecology*, **48**, 501–7.

Egglishaw, H.J., Gardiner, W.R., Shackley, P.E. & Struthers, G. (1984) Principles and practice of stocking streams with salmon eggs and fry. *Department of Agriculture and Fisheries for Scotland. Scottish Fisheries Information pamphlet*, **10**, 1–22.

Egglishaw, H.J. & Shackley, P.E. (1973) An experiment on faster growth of salmon *Salmo salar* L. in a Scottish stream. *Journal of Fish Biology*, **5**, 197–204.

Egglishaw, H.J. & Shackley, P.E. (1982) Influence of water depth on dispersion of juvenile salmonids, *Salmo salar* L. and trout, *S. trutta* L,. in a Scottish stream. *Journal of Fish Biology*, **21**, 141–55.

Egglishaw, H.J. & Shackley, P.E. (1985) Factors governing the production of juvenile Atlantic salmon in Scottish streams. *Journal of Fish Biology*, **27 (A)**, 27–33.

Einum, S. & Fleming, I.A. (1997) Genetic divergence and interactions in the wild among native, farmed and hybrid Atlantic salmon. *Journal of Fish Biology*, **50**, 634–51.

Elliott, J.M. (1975a) The growth rate of brown trout, *Salmo trutta* L., fed on maximum rations. *Journal of Animal Ecology*, **44**, 805–21.

Elliott, J.M. (1975b) The growth rate of brown trout, *Salmo trutta* L., fed on reduced rations. *Journal of Animal Ecology*, **44**, 823–42.

Elliott, J.M. (1975c) Number of meals in a day, maximum weight of food consumed in a day and maximum rate of feeding for brown trout, *Salmo trutta* L., *Freshwater Biology*, **5**, 287–303.

Elliott, J.M. (1976a) The downstream drifting of eggs of brown trout *Salmo trutta* L. *Journal of Fish Biology*, **9**, 45–50.

Elliott, J.M. (1976b) The energetics of feeding, metabolism and growth of brown trout, *Salmo trutta* L., in relation to body weight, water temperature and ration size. *Journal of Animal Ecology*, **45**, 923–48.

Elliott, J.M. (1976c) Energy losses in the waste products of brown trout (*Salmo trutta* L.). *Journal of Animal Ecology*, **45**, 561–80.

Elliott, J.M. (1981) Some aspects of thermal stress in freshwater teleosts. In: *Stress and Fish* (ed. A.D. Pickering). Academic Press, London, 209–45.

Elliott, J.M. (1984a) Numerical changes and population regulation in young migratory trout, *Salmo trutta*, in a Lake District stream, 1966–83. *Journal of Animal Ecology*, **53**, 327–50.

Elliott, J.M. (1984b) Growth, size, biomass and production of young migratory trout *Salmo trutta* in a Lake District stream, 1966–83. *Journal of Animal Ecology*, **53**, 979–94.

Elliott, J.M. (1986) Spatial distribution and behavioural movements of migratory trout (*Salmo trutta*) in a Lake District stream. *Journal of Animal Ecology*, **55**, 907–22.

Elliott, J.M. (1991) Tolerance and resistance to thermal stress in juvenile Atlantic salmon, *Salmo salar*. *Freshwater Biology*, **25**, 61–70.

Elliott, J.M. (1992) Analysis of sea trout catch statistics for England and Wales. *Fisheries Technical Report. National Rivers Authority*, **2**, 1–46.

Elliott, J.M. (1994) *Quantitative Ecology and the Brown Trout*. Oxford University Press, Oxford.

Elliott, J.M. (1997) Stomach contents of adult sea trout caught in six English rivers. *Journal of Fish Biology*, **50**, 1129–32.

Elliott, J.M. & Hurley, M.A. (1997) A functional model for maximum growth of Atlantic salmon parr, *Salmo salar*, from two populations in northwest England. *Functional Ecology*, **11**, 592–603.

Elliott, J.M. & Hurley, M.A. (1998a) Population regulation in adult, but not juvenile, resident trout (*Salmo trutta*) in a Lake District stream. *Journal of Animal Ecology*, **67**, 280–6.

Elliott, J.M. & Hurley, M.A. (1998b) An individual-based model for predicting the emergence period of sea trout fry in a Lake District stream. *Journal of Fish Biology*, **53**, 414–33.

Elliott, J.M., Hurley, M.A. & Fryer, R.J.C. (1995) A new improved growth model for brown trout, *Salmo trutta*. *Functional Ecology*, **9**, 290–8.

Elson, P.F. (1957) The importance of size in the change from parr to smolt in Atlantic salmon. *Canadian Fish Culturist*, **21**, 1–6.

Elson, P.F. & Tuomi, A.L.W. (1975) The Foyle Fisheries: new bases for rational management. *Special Report to the Foyle Fisheries Commission, Londonderry, Northern Ireland* 1–224.

Engstrom-Heg, R. (1986) Prediction of wild brown trout catch rates from established yearling population density and fishing intensity. *North American Journal of Fisheries Management*, **6**, 410–17.

Environment Agency (1997a) Frome and Piddle catchment management plan. *First Annual Review*. Environment Agency, South West Region.

Environment Agency (1997b) *Understanding buffer zones*. Environment Agency.

Environment Agency (1997c) *Salmonid and Freshwater Fisheries Statistics for England and Wales, 1995.* Environment Agency, Bristol.

Environment Agency (1998a) *A price worth paying. The Environment Agency's Programme for Water Companies 2000–2005.* Environment Agency, Bristol.

Environment Agency (1998b) *Fisheries habitat improvement.* Environment Agency.

Environment Agency (1998c) Diffuse pollution from agriculture a field guide. *Environment Agency R & D Publication*, **13**, 1–17.

Environment Agency (1998d) *Understanding riverbank erosion from a conservation perspective.* Environment Agency, 1–19.

Environment Agency (1999) *Understanding rural land use.* Environment Agency.

Evans, E.C., Greenwood, M.T. & Petts, G.E. (1995) Thermal profiles within river beds. *Hydrological Processes*, **9**, 19–25.

Evans, E.C. & Petts, G.E. (1997) Hyporheic temperature patterns within riffles. *Hydrological Sciences Journal*, **42**, 199–213.

Everall, N.C., MacFarlane, N.A.A. & Sedgwick, R.W. (1989) The interactions of water hardness and pH with the acute toxicity of zinc to brown trout, *Salmo trutta* L. *Journal of Fish Biology*, **35**, 27–36.

Fahy, E. (1983) Food and gut parasite burden of migratory trout *Salmo trutta* L. in the sea. *Irish Naturalists' Journal*, **21**, 11–18.

Fahy, E. (1985) Feeding, growth and parasites of trout *Salmo trutta* L. from Mulroy Bay, an Irish sea lough. *Irish Fishery Investigations*, **25 (A)**, 1–12.

Fernö, A. & Järvi, T. (1998) Domestication genetically alters the anti-predator behaviour of anadromous Brown Trout (*Salmo trutta*) – a dummy predator experiment. *Nordic Journal of Freshwater Research*, **74**, 95–100.

Finnigan, R.J., Marshall, D.E., Mundie, J.H., Slaney, P.A. & Taylor, G.D. (1980) *Stream Enhancement Guide.* Fisheries and Oceans, Vancouver, British Columbia, 1–95.

Fleming-Jones, D. & Stent, R.F. (1975) Factors affecting the trout fishery at Grafham Water. *Fishery Management*, **6**, 558–63.

Fluskey, R.D. (1989) An analysis of the gravels used by spawning salmonids. *Irish Fishery Investigations Series A (Freshwater)*, **32**, 1–14.

Forest Industry Council of Great Britain (1996) *Yearbook.* Forest Industry Council of Great Britain.

Forestry Commission (1993) *Forest and Water Guidelines.* Forestry Commission.

Forsythe, M.G. (1968) Analysis of the 1966 smolt run in the North-West Miramichi River, New Brunswick. *Fisheries Research Board of Canada Technical Report*, **91**, 1–33.

Fraser, J.C. (1975) Determining discharges for fluvial resources. *F.A.O. Fisheries Technical Paper*, **143**, 1–102.

Fraser, N.H.C., Heggenes, J., Metcalfe, N.B. & Thorpe, J.E. (1995) Low summer temperatures cause juvenile Atlantic salmon to become nocturnal. *Canadian Journal of Zoology*, **73**, 446–51.

Frost, W.E. (1974) A survey of the rainbow trout (*Salmo gairdneri*) in Britain and Ireland. *Salmon and Trout Association*, 1–36.

Frost, W.E & Brown, M.E. (1967) *The Trout.* Collins, London.

Gangmark, H.A. & Bakkala, R.G. (1960) A comparative study of unstable and stable (artificial channel) spawning streams for incubating King salmon at Mill Creek. *California Fish and Game*, **46**, 151–64.

Gardiner, W.R. & Geddes, P. (1980) The influence of body composition on the survival of juvenile salmon. *Hydrobiogia*, **69**, 67–72.

Gardiner, W.R. & Shackley, P. (1991) Stock and recruitment and inversely density-dependent growth of salmon, *Salmo salar* L. in a Scottish Stream. *Journal of Fish Biology*, **38**, 691–6.

Gardner, M.L.G. (1971) Recent changes in the movements of adult salmon *Salmo salar* L. in the Tory-Tummel-Garry system. *Journal of Fish Biology*, **3**, 83–95.

Garner, P., Clough, S., Griffiths, S.W., Deans, D. & Ibbotson, A. (1998) Use of shallow marginal habitat by *Phoxinus phoxinus*: a trade-off between temperature and food? *Journal of Fish Biology*, **52**, 600–609.

Garside, E.T. (1959) Some effects of oxygen in relation to temperature on the development of embryos of brook trout and rainbow trout. *Canadian Journal of Zoology*, **37**, 489–698.

Garside, E.T. (1966) Effects of oxygen in relation to temperature on the development of lake trout embryos. *Journal of the Fisheries Research Board of Canada*, **23**, 1037–134.

Gee, A.S. & Milner, N.J. (1980) Analysis of 70-year catch statistics for Atlantic salmon (*Salmo salar*) in the River Wye and its implications for management of stocks. *Journal of Applied Ecology*, **17**, 41–57.

Gee, A.S., Milner, N.J. & Hemsworth, R.J. (1978) The effect of density on mortality in juvenile Atlantic salmon (*Salmo salar*). *Journal of Animal Ecology*, **47**, 497–505.

Geertz-Hansen, P. & Mortensen, E. (1983) The effect of dissolved and precipitated iron on the reproduction of brown trout (*Salmo trutta*). *Vatten*, **39**, 55–62.

Gibson, R.J. (1973) Interactions of juvenile Atlantic salmon (*Salmo salar* L.) and brook trout (*Salvelinus fontinalis* Mitchill). *Special Publication of the International Atlantic Salmon Foundation Series 3*, **4**, 181–202.

Gibson, R.J. (1988) Mechanisms regulating species composition, population structure and production of stream salmonids: a review. *Polski Archiwum Hydrobiologii*, **35**, 469–95.

Gibson, R.J. (1993) The Atlantic salmon in fresh water: spawning, rearing and production. *Reviews in Fish Biology and Fisheries*, **3**, 39–73.

Giles, N. & Associates (1998) Freshwater Fisheries and Wildlife Conservation a good practice guide. *Environment Agency*, 1–37.

Giles, N., Phillips, B. & Barnard, S. (1991) Ecological effects of low flows in chalk streams. *The Game Conservancy Review of 1991*, 90–1.

Glover, R.D. (1986) Trout stream rehabilitation in the Black Hills of South Dakota. In: *5th Trout Stream Habitat Improvement Workshop* (eds J.G. Miller, J.A. Arway & R.F. Carline). Lock Haven University, 7–15.

Goody, N.P. (1988) The hydrology of the River Spey. In: *Land use in the River Spey catchment* (ed. D. Jenkins). ACLU Symposium Series No. 1. Aberdeen Reminder Press.

Gore, J.A. (1985) *The Restoration of Rivers and Streams: Theories and Experience.* Butterworths, London.

Gorham, E. (1957) The chemical composition of rain from Rosscahill in County Galway. *Irish Naturalists' Journal*, **XII**, 1–4.

Gorham, E. (1958a) The influence and importance of daily weather conditions in the supply of chloride, sulphate and other ions to fresh waters from atmospheric precipitation. *Philosophical Transactions of the Royal Society of London, Series B*, **241**, 147–78.

Gorham, E. (1958b) Atmospheric pollution by hydrochloric acid. *Quarterly Journal of the Royal Meteorological Society*, **84**, 274–6.

Grant, J.W.A. & Kramer, D.L (1990) Territory size as a predictor of the upper limit to population density of juvenile salmonids in streams. *Canadian Journal of Fisheries and Aquatic Sciences*, **47**, 1724–37.

Gray, J. (1928) The growth of fish. III. The effects of temperature on the development of eggs of *Salmo fario*. *British Journal of Experimental Biology*, **6**, 110–24.

Gray, J.R.A. & Edington, J.M. (1969) Effect of woodland clearance on stream temperature. *Journal of the Fisheries Research Board of Canada*, **26**, 399–403.

Greeley, J.R. (1932) The spawning habits of brook, brown and rainbow trout and the problems of egg predators. *Transactions of the American Fisheries Society*, **62**, 239–48.

Gregory, J. (1989) *Water Schemes – The Safeguarding of Fisheries*. Atlantic Salmon Trust.

Gregory, S.V. & Bisson, P.A. (1997) Degradation and loss of anadromous salmonid habitat in the Pacific northwest. In: *Pacific Salmon and their ecosystems* (eds D.J. Stouder, P.A. Bisson & R.J. Naiman). Chapman & Hall, New York, 277–314.

Griffith, J.S. & Smith, R.W. (1993) Use of winter concealment cover by juvenile cutthroat and brown trout in the south fork of the Snake River, Idaho. *North American Journal of Fisheries Management*, **13**, 823–30.

Grost, R.T., Hubert, W.A. & Wesche, T.A. (1991) Description of brown trout redds in a mountain stream. *Transactions of the American Fisheries Society*, **120**, 582–8.

Gunnes, K. (1979) Survival and development of Atlantic salmon eggs and fry at three different temperatures. *Aquaculture*, **16**, 211–18.

Gustard, A., Bullock, A. & Dixon, J.M. (1992) Low flow estimation in the United Kingdom. *Institute of Hydrology*, Report 108.

Gustard, A. & Elliott, C.R.N. (1997) The role of hydro-ecological models in the development of sustainable water resources. *Sustainability of Water Resources under Increasing Uncertainty* (Proceedings of the Rabat Symposium S1, April 1997). IAHS Publication, **240**, 407–17.

Hamor, T. & Garside, E.T. (1975) Regulation of oxygen consumption by incident illumination of embryonated ova of Atlantic salmon, *Salmo salar* L. *Comparative Biochemistry and Physiology, Series A*, **52**, 277–80.

Hamor, T. & Garside, E.T. (1976) Developmental rates of Atlantic salmon, *Salmo salar* L., in response to various levels of temperature, dissolved oxygen and water exchange. *Canadian Journal of Zoology*, **54**, 1912–17.

Hamor, T. & Garside, E.T. (1977) Size relations and yolk sac utilisation in embryonated ova and alevins of Atlantic salmon *Salmo salar* L. in various combinations of temperature and dissolved oxygen. *Canadian Journal of Zoology*, **55**, 1892–8.

Hansen, L.P. & Jonsson, B. (1991) Evidence of a genetic component in the seasonal return pattern of Atlantic salmon, *Salmo salar* L. *Journal of Fish Biology*, **38**, 251–8.

Hardy, J.C. (1963) An examination of eleven stranded redds of brown trout (*Salmo trutta*) excavated on the Selwyn River during July and August 1960. *New Zealand Journal of Science*, **6**, 107–19.

Hardy, J.C. (1972) South Island Council of acclimatisation societies. Proceedings of the Quinnat Salmon fishery Symposium 2–3 October 1971 – Ashburton. *New Zealand Ministry of Agriculture and Fisheries, Fisheries Technical Report*, **83**, 1–298.

Harper, D.M. & Ferguson, A.J.D. (1995) (eds) *The Ecological Basis for River Management*. John Wiley & Sons, Chichester.

Hartman, G.F., Scrivener, J.C. & Miles, M.J. (1994) Impacts of logging in Carnation Creek, a high-energy coastal stream in British Columbia and their implication for restoring fish habitat. *Canadian Journal of Fisheries and Aquatic Sciences*, **53**, **(1)**, 237–51.

Hawkes, S.P. (1978) Stranded redds of chinook salmon in the Mathias River, South Island, New Zealand. *New Zealand Journal of Marine and Freshwater Research*, **12**, 167–71.

Hayes, F.R. (1953) Artificial freshets and other factors controlling the ascent and population of Atlantic salmon in the Le Havre River N.S. *Bulletin of the Fisheries Board of Canada*, **99**, 1–47.

Hayes, F.R., Wilmot, I.R. & Livingstone, D.A. (1951) The oxygen consumption of the salmon egg in relation to development and activity. *Journal of Experimental Biology*, **116**, 377–95.

Heard, W.R. (1991) Life history of the pink salmon (*Oncorhynchus gorbuscha*). In: *Pacific Salmon Life Histories* (eds C. Groot & L. Margolis). Vancouver, U.B.C. Press, 119–230.

Heggberget, T.G. (1988) Timing of spawning in Norwegian Atlantic salmon (*Salmo salar*). *Canadian Journal of Fisheries and Aquatic Sciences*, **45**, 845–9.

Heggberget, T.G. (1991) Some environmental requirements of Atlantic salmon. *American Fisheries Society Symposium*, **10**, 132–5.

Heggberget, T.G., Haukebø, T., Mork, J. & Staal, G. (1988) Temporal and spatial segregation of spawning in sympatric populations of Atlantic salmon, *Salmo salar* L., and brown trout, *S. trutta* L., *Journal of Fish Biology*, **33**, 347–56.

Heggenes, J. (1988) Physical habitat selection by brown trout (*Salmo trutta*) in riverine systems. *Nordic Journal of Freshwater Research*, **64**, 74–90.

Heggenes, J. & Borgstrøm, R. (1991) Effect of habitat types on survival, spatial distribution and production of an allopatric cohort of Atlantic salmon, *Salmo salar* L., under conditions of low competition. *Journal of Fish Biology*, **38**, 267–80.

Heggenes, J., Krog, O.M.W., Lindås, O.R., Dokk, J.G. & Bremnes, T. (1993) Homeostatic behavioural responses in a changing environment: brown trout (*Salmo trutta*) become nocturnal during winter. *Journal of Animal Ecology*, **62**, 295–308.

Heggenes, J. & Traaen, T. (1988) Downstream migration and initial water velocities in stream channels, for fry of four salmonid species. *Journal of Fish Biology*, **32**, 717–27.

Hellawell, J.M. (1976) River Management and the migratory behaviour of salmonids. *Fishery Management*, **7**, 57–60.

Hellawell, J.M. (1986) *Biological Indicators of Freshwater Pollution and Environmental Management*. Elsevier Applied Science Publishers, London & New York.

Hellawell, J.M., Leatham, H. & Williams, G.I. (1974) The upstream migratory behaviour of salmonids in the River Frome, Dorset. *Journal of Fish Biology*, **6**, 729–44.

Herricks, E.E. & Osborne, L.L (1985) Water quality restoration and production in streams and rivers. In: *The Restoration of Rivers and Streams: Theories and Experience* (ed. J.A. Gore). Butterworth, Boston, USA, 1–20.

Hoffman, C.C., Pedersen, M.L. Kronvang, B. & Øvig, L. (1998) Restoration of the Rivers Brede, Cole and Skerne: Joint Danish and British EU-LIFE demonstration project: IV – Implications for Nitrate and Iron Transformation. *Aquatic Conservation Marine and Freshwater Ecosystems*, **8**, 223–40.

Holcik, J. (1990) Conservation of the huchen, *Hucho hucho* (L.) (Salmonidae) with special reference to Slovakian rivers. *Journal of Fish Biology*, **37 (A)**, 113–21.

Holcik, J., Hensel, K., Nielslanik, J. & Skacel, L. (1988) The Eurasian Huchen, *Hucho hucho*, largest salmon in the world. In: *Perspectives in Vertebrate Science* (ed. E.K. Balon). Junk, Dordrecht, Boston, Lancaster, 1–239.

Holden, A.V. (1988) *The automatic counter – a tool for the management of fisheries.* Atlantic Salmon Trust, 1–36.

Holmes, N.T.H. & Nielsen, M.B. (1998) Restoration of the Rivers Brede, Cole and Skerne: Joint Danish and British EU-LIFE Demonstration Project: 1 – Setting up and delivery of the Project. *Aquatic Conservation Marine and Freshwater Ecosystems*, **8**, 185–96.

Holtby, L.B. & Healey, M.C. (1986) Selection for adult size in female coho salmon (*Oncorhynchus kisutch*). *Canadian Journal of Aquatic Sciences*, **43**, 1946–59.

Humpesch, U.H. (1985) Inter– and intra-specific variation in hatching success and embryonic development of five species of salmonids and *Thymallus thymallus*. *Archiv für Hydrobiologie*, **104**, 129–44.

Hunt, R.L. (1976) A long term evaluation of trout habitat development and its relation to improved management-related research. *Transactions of the American Fisheries Society*, **105**, 361–4.

Hunt, R.L. (1988) A compendium of 45 trout stream habitat development evaluations in Wisconsin during 1953–1985. *Technical Bulletin Department of Natural Resources, Wisconsin*, **162**, 1–80.

Hunter, C.J. (1991) *Better Trout Habitat: A Guide to Restoration and Management.* Washington: Island Press, 1–320.

Huntingford, F.A. & De Leaniz, C.G. (1997) Social dominance and the acquisition of profitable feeding sites in juvenile Atlantic salmon. *Journal of Fish Biology*, **51**, 1009–14.

Huntingford, F.A., Metcalfe, N.B. & Thorpe, J.E. (1988) Choice of feeding station in Atlantic salmon, *Salmo salar*, parr: effects of predation risk, season and life history strategy. *Journal of Fish Biology*, **33**, 917–24.

Huntsman, A.G. (1948) Freshets and fish. *Transactions of the American Fisheries Society*, **75**, 257–66.

Hvidsten, N.A., Heggberget, T.G. & Hansen, L.P. (1994) Homing and straying of hatchery-reared Atlantic salmon, *Salmo salar* L., released in three rivers in Norway. *Aquaculture and Fisheries Management*, **25**, 9–16.

Hvidsten, N.A., Jensen, A.J., Vivås, H., Bakke, O. & Heggberget, T.G. (1995) Downstream migration of Atlantic salmon smolts in relation to water flow, water temperature, moon phase and social interaction. *Nordic Journal of Freshwater Research*, **70**, 38–48.

Ievleva, M.Y. (1967) Resistance of developing eggs of sockeye salmon to mechanical action. *Isvestiya Tikhookeanskogo i Okeanografii (Vladivostok)*, **57**, 555–79. (Canadian Translation of Fisheries and Aquatic Sciences 4617, 1980.)

Institute of Hydrology (1975) *Flood Studies Report.* Natural Environment Research Council.

Irvine, J. R. (1986) Effects of varying discharge on the downstream movement of salmon fry, *Oncorhynchus tshawytscha* Walbaum. *Journal of Fish Biology*, **28**, 17–28.

Jenkins, T.M. (1969) Social structure, position choice and microdistribution of two trout species (*Salmo trutta* and *Salmo gairdneri*) resident in mountain streams. *Animal Behaviour Monographs*, **2**, 57–123.

Jensen, A.J. (1990) Growth of young migratory brown trout *Salmo trutta* correlated with water temperature in Norwegian rivers. *Journal of Animal Ecology*, **59**, 603–14.

Jensen, J.O.T. & Alderdice, D.F. (1983) Changes in mechanical shock sensitivity of coho salmon (*Oncorhynchus kisutch*) eggs during incubation. *Aquaculture*, **32**, 303–12.

Johnsen, B.O. & Jensen, A.J. (1986) Infestations of Atlantic salmon, *Salmo salar*, by *Gyrodactylus salaris* in Norwegian rivers. *Journal of Fish Biology*, **29**, 233–41.

Jones, J.W. (1959) *The Salmon*. Collins, London, 1–192.

Jones, J.W. & Ball, J.N. (1954) The spawning behaviour of brown trout and salmon. *Journal of Animal Behaviour*, **2**, 103–14.

Jones, J.W. & King, G.M. (1949) Experimental observations on the spawning behaviour of the Atlantic salmon (*Salmo salar* Linn.). *Proceedings of the Zoological Society, London*, **119**, 33–48.

Jones, J.W. & King, G.M. (1950) Further experimental observations on the spawning behaviour of Atlantic salmon (*Salmo salar* Linn.). *Proceedings of the Zoological Society, London*, **120**, 317–23.

Jonsson, B. (1985) Life history patterns of freshwater resident and sea–run brown trout in Norway. *Transactions of the American Fisheries Journal*, **114**, 182–94.

Jonsson, B. (1989) Life history and habitat use of Norwegian brown trout (*Salmo trutta*). *Freshwater Biology*, **21**, 71–86.

Jowett, I.G. (1990) Factors related to the distribution and abundance of brown and rainbow trout in New Zealand clear–water rivers. *New Zealand Journal of Marine and Freshwater Research*, **24**, 429–40.

Jungwirth, M. & Winkler, H. (1984) The temperature dependence of embryonic development of grayling (*Thymallus thymallus*), Danube salmon (*Hucho hucho*), arctic char (*Salvelinus alpinus*) and brown trout (*Salmo trutta fario*). *Aquaculture*, **38**, 315–27.

Kalleberg, H. (1958) Observations in a stream tank of territoriality and competition in juvenile salmon and trout (*Salmo salar* L. and *S. trutta* L.). *Report of the Institute of Freshwater Research, Drottningolm*, **39**, 55–98.

Kemp, S. (1986) A helping hand. *Salmon and Trout Magazine*, **232**, 43–5.

Kennedy, G.J.A. (1982) Factors affecting the survival and distribution of salmon (*Salmo salar* L.) stocked in upland trout (*Salmo trutta* L.) streams in Northern Ireland. *E.I.F.A.C. Technical Paper*, **42**, 227–42.

Kennedy, G.J.A. & Crozier, W.W. (1995) Factors affecting recruitment success in salmonids. In: *The Ecological Basis for River Management* (eds D.M. Harper & A.J.D. Ferguson). John Wiley & Sons, Chichester, 349–62.

Kennedy, G.J.A. & Strange, C.D. (1982) The distribution of salmonids in upland streams in relation to depth and gradient. *Journal of Fish Biology*, **20**, 579–91.

Kirby, C., Newson, M.D. & Gilman, K. (1991) Plynlimon Research: The first two decades. *Institute of Hydrology, Wallingford*, Report **109**, 1–188.

Kocik, J.F. & Taylor, W.W. (1994) Summer survival and growth of brown trout with and without steelhead under equal total salmonine densities in an artificial stream. *Transaction of the American Fisheries Society*, **123**, 931–8.

Kondolf, G.M. (1998) Post–project evaluation of river and stream restoration. In: *United*

Kingdom Floodplains (eds R.G. Bailey, P.V. José & B.R. Sherwood). Westbury Academic & Scientific Press, Otley, 465–73.

Kondolf, G.M., Coda, G.F. & Sale, M.J. (1987) Assessing flushing-flow requirements for brown trout spawning gravels in steep streams. *Water Resources Bulletin*, **23**, 927–35.

Kondolf, G.M., Sale, M.J. & Wolman, M.G. (1993) Modification of fluvial gravel size by spawning salmonids. *Water Resources Research*, **29**, 2265–74.

Kondolf, G.M. & Wolman, M.G. (1993) The size of salmonid spawning gravels. *Water Resources Research*, **29**, 2275–85.

Kronvang, B., Svendsen, L.M., Brookes, A., Fisher, K., Møller, B., Ottosen, O., Newson, M. & Sear, D. (1998) Restoration of the Rivers Brede, Cole and Skerne: Joint Danish and British EU-LIFE Demonstration Project, III – Channel morphology, hydrodynamics and transport of sediment and nutrients. *Aquatic Conservation Marine and Freshwater Ecosystems*, **8**, 209–22.

L'Abée-Lund, J.H. & Hindar, K. (1990) Interpopulation variation in reproductive traits of anadromous female brown trout *Salmo trutta* L. *Journal of Fish Biology*, **37**, 755–63.

Lamond, H. (1916) *The sea trout*. Sherrat & Hughes, London and Manchester, 1–219.

Le Cren, E.D. (1973) Population dynamics of salmonid fish. In: *The Mathematical Theory of the Dynamics of Biological Populations* (eds M.S. Bartlett & R.W. Hiarns). Academic Press, London, 125–35.

Le Cren, E.D. (1985) *The biology of the sea trout*. Atlantic Salmon Trust, 1–42.

Lesel, R., Therezien, Y. & Vibert, R. (1971) Introduction des Salmonidés aux Iles Kerguelen. I – Premiers résultats et observations préliminaires. *Annales d'Hydrobiologie*, **2**, 275-304.

Lindroth, A.C. (1942) Sauerstoffverbrough der Fische. II Verschiedene Entwicklungs und altersstadien vom Lachs und Hecht. *Zeitschrift für vergleichende Physiologie*, **29**, 585–94.

Lister, D.B. & Walker, C.E. (1966) The effect of flow control on freshwater survival of chum, coho and chinook salmon in Big Qualicum River. *Canadian Fish Culture*, **40**, 41–9.

Lowe, S. (1996) *Fish habitat enhancement designs: typical structures*. Alberta Environmental Protection, Water Resources Management Services, River Engineering Branch, Edmonton, Alberta.

MacCrimmon, H.R. (1954) Stream studies on planted Atlantic salmon. *Journal of the Fisheries Research Board of Canada*, **11**, 362–403.

MacCrimmon, H.R. & Gots, B.A. (1986) Laboratory observations on emergence patterns of juvenile Atlantic salmon, *Salmo salar*, relative to sediment loadings of test substrate. *Canadian Journal of Zoology*, **64**, 1331–6.

MacCrimmon, H.R. & Marshall, T.L. (1968) World distribution of Brown Trout, *Salmo trutta*. *Journal of the Fisheries Research Board of Canada*, **25**, 2527–48.

MacCrimmon, H.R., Marshall, T.L. & Gots, B.L. (1970) World Distribution of Brown Trout, *Salmo trutta*: further observations. *Journal of the Fisheries Research Board of Canada*, **27**, 811–18.

Maitland, P.S. (1972) Key to British freshwater fishes. *Freshwater Biological Association Scientific Publication*, **27**, 1–139.

Mann, R.H.K. & Aprahamian, M.W. (1996) Fish Pass Technology Training Course. *Institute of Freshwater Ecology & Environment Agency*, 1–45.

Mann, R.H.K., Blackburn, J.H. & Beaumont, W.R.C. (1989) The ecology of brown trout *Salmo trutta* in English chalk streams. *Freshwater Biology*, **21**, 57–70.

Mann, R.H.K. & Winfield, I.J. (1992) Restoration of Riverine Fisheries Habitat. *National Rivers Authority R & D Note*, **105**, 1–56.

Marschall, E.A. & Crowder, L.B. (1995) Density-dependent survival as a function of size in juvenile salmonids in streams. *Canadian Journal of Aquatic Sciences*, **52**, 136–40.

Marty, C. & Beall, E. (1989) Modalités spatio-temporelles de la dispersion d'alevins de saumon atlantique (*Salmo salar* L.) à l'émergence. *Revue des Sciences de l'eau*, **2**, 831–46.

Marty, C., Beall, E. & Gots, B.A. (1986) Influence de quelques parametres du milieu d'incubation sur la survie d'alevins de saumon atlantique, *Salmo salar* L., en ruisseau experimental. *Internationale Revue der Gesamten Hydrobiologie und Hydrographie*, **71**, 349–61.

Mason, J.C. (1969) Hypoxial stress prior to emergence and competition among coho salmon. *Journal of the Fisheries Research Board of Canada*, **26**, 63–91.

Maughan, O.E., Nelson, K.L. & Ney, J.J. (1978) Evaluation of stream improvement practices in southeastern trout streams. *Virginia Water Resources Research Center. Publication of Virginia Polytechnic Institute and State University, Blacksburg, Virginia Bulletin*, **115**, 1–67.

McPhail, J.D. (1997) The origin and speciation of *Oncorhynchus* revisited. In: *Pacific salmon & their ecosystems* (eds D.J. Stouder, P.A. Bisson & R.J. Naiman). Chapman & Hall, New York, 29–38.

McWilliam, P.G. (1982) A comparison of the physiological characteristics in normal and acid exposed populations of brown trout *Salmo trutta. Comparative Biochemistry and Physiology Series A*, **73**, 515–22.

Meek, E.M. (1925) The pollution of the River Tyne. *Report of the Dove Marine Laboratory for the year ending June 30th, 1925*.

Metcalfe, N.B., Huntingford, F.A., Graham, W.D. & Thorpe, J.E. (1989) Early social status and development of life-history strategies in Atlantic salmon. *Proceedings of the Royal Society of London*, **236**, 7–19.

Metcalfe, N.B., Huntingford, F.A., Thorpe, J.E. & Adams, C.E. (1990) The effects of social status on life-history variation in juvenile salmon. *Canadian Journal of Zoology*. **68**, 2630–6.

Metcalfe, N.B. & Thorpe, J.E. (1990) Determinants of geographical variation in age of seaward migrating salmon, *Salmo salar. Journal of Animal Ecology*, **59**, 135–45.

Metcalfe, N.B. & Thorpe, J.E. (1992) Anorexia and defended energy levels in over-wintering juvenile salmonids. *Journal of Animal Ecology*, **61**, 175–81.

Mills, D.H. (1971) *Salmon and trout: A resource, its ecology, conservation and management*. Oliver & Boyd, Edinburgh.

Mills, D.H. (1989) *Ecology and Management of Atlantic Salmon*. Chapman & Hall, London and New York.

Mills, D.H. (1991) *Strategies for the Rehabilitation of Salmon Rivers*. Atlantic Salmon Trust, Institute of Fisheries Management, Linnaen Society London.

Mills, D.H. (1993) *Salmon in the sea and new enhancement strategies*. Fishing News Books, Oxford.

Milner, N.J., Hemsworth, R.J. & Jones, B.E. (1985) Habitat evaluation as a management tool, *Journal of Fish Biology*, **27 (A)**, 85–108.

Milner, N.J., Scullion, J., Carling, P.A. & Crisp, D.T. (1981) The effects of discharge on sediment dynamics and consequent effects on invertebrates and salmonids in upland waters. *Advances in Applied Biology*, **6**, 153–220.

Ministry of Agriculture, Fisheries and Food & Welsh Office Agriculture Department (1998a) *Code of good agricultural practice for the protection of the soil*. Ministry of Agriculture, Fisheries and Food & Welsh Office Agriculture Department.

Ministry of Agriculture, Fisheries and Food & Welsh Office Agriculture Department (1998b) *Code of good agricultural practice for the protection of water*. Ministry of Agriculture, Fisheries and Food & Welsh Office Agriculture Department.

Moir, H.J., Soulsby, C. & Youngson, A. (1998) Hydraulic and sedimentary characteristics of habitat utilized by Atlantic salmon for spawning in Girnock Burn, Scotland. *Fisheries Management and Ecology*, **5**, 241–54.

Montgomery, F.A.C., Thom, N.S. & Cockburn, A. (1964) Determination of dissolved oxygen by the Winkler method and the solubility of oxygen in pure water and sea water. *Journal of Applied Chemistry*, **14**, 280–96.

Moore, A. & Scott, A. (1988) Observations on recently emerged sea trout, *Salmo trutta* L., fry in a chalk stream using a low-light underwater camera. *Journal of Fish Biology*, **33**, 959–69.

Mundie, J.H. & Bell-Irving, K.H. (1986) Predictability of the consequences of the Kemono-Hydroelectric proposal for natural salmon populations. *Canadian Water Resources Journal*, **11**, 14–25.

Mundie, J.H. & Crabtree, D.G. (1997) Effects on sediments and biota of cleaning a salmonid spawning channel. *Fisheries Management and Ecology*, **4**, 111–26.

Mundie, J.H., Mounce, D.E. & Simpson, K.S. (1990) Semi-natural rearing of coho salmon, *Oncorhynchus kisutch* (Walbaum), smolts, with an assessment of survival to the catch and escapement. *Aquaculture and Fisheries Management*, **21**, 327–45.

Munro, N.R. & Balmain, K.H. (1956) Observations on the spawning runs of brown trout in the South Quiech, Loch Leven. *Freshwater and Salmon Fisheries Research*, **13**, 1–17.

Munro, W.R. (1965) Effects of passage through hydro-electric turbines on salmonids. *ICES CM 1965 Salmon and trout communication*, **57**, 1–4.

Myers, R.A. & Hutchings, J.A. (1987) Mating of anadromous Atlantic salmon, *Salmo salar* L., with mature male parr. *Journal of Fish Biology*, **31**, 143–6.

Nagata, M. & Irvine, J.R. (1997) Differential dispersal patterns of male and female masu salmon fry. *Journal of Fish Biology*, **51**, 601–6.

NASCO (1996) Report of the ICES Advisory Committee on Fishery Management. *NASCO Report of the Annual Meeting of the Council 1996 Thirteenth Annual Meeting, Gothenburg, Sweden*. 81–124.

Naslund, I. (1987) Effects of habitat improvement on the brown trout (*Salmo trutta* L.) population of a north Swedish stream. *Report of the Institute of Freshwater Research, Drottningholm*, 1–28.

National Research Council (1992) *Restoration of Aquatic Ecosystems, Science, Technology and Public Policy*. National Academy Press, Washington, D.C.

National Research Council (1996) *Upstream: Salmon and Society in the Pacific Northwest*. National Academy Press, Washington, D.C.

National Rivers Authority (1993) *Salmonid and Freshwater Fisheries Statistics for England and Wales, 1991. National Rivers Authority, Bristol.*

National Rivers Authority (1994a) *National Angling Survey* 1994. HMSO, London.

National Rivers Authority (1994b) *Guidance notes for local planning authorities on methods of protecting the water environment through development plans.* National Rivers Authority.

National Rivers Authority (1994c) *Salmonid and Freshwater Fisheries Statistics for England and Wales, 1992.* National Rivers Authority, Bristol.

National Rivers Authority (1994d) *Salmonid and Freshwater Fisheries Statistics for England and Wales, 1993.* National Rivers Authority, Bristol.

National Rivers Authority (1995) *Salmonid and Freshwater Fisheries Statistics for England and Wales, 1994.* National Rivers Authority.

Neal, C., Smith, C.J. & Hill, S. (1992) Forestry impact on upland waters. *Institute of Hydrology Report*, **119**, 1–50.

Neel, J.K. (1963) Impact of reservoirs. In: *Limnology in North America* (ed. D.G. Frey). University of Winsconsin Press, 573–93.

Newman, E.I. (1997) Phosphorus balance of contrasting farming systems, past and present. Can food production be sustainable? *Journal of Applied Ecology*, **34**, 1334–47.

Newson, M.D. (1981) Mountain streams. In: *British Rivers* (ed. J. Lewin). George Allen & Unwin, London.

Newson, M.D. (1994) *Hydrology and the River Environment.* Oxford University Press.

Nicieza, A.G. & Brana, F. (1993) Relationships among smolt size, marine growth, and sea age at maturity of Atlantic salmon (*Salmo salar*) in northern Spain. *Canadian Journal of Fisheries and Aquatic Sciences*, **50**, 1632–40.

North, E. (1983) Relationship between stocking and anglers' catches at Draycote Water trout fishery. *Fishery Management*, **14**, 187–98.

Northcote, T.G. (1997) Protandromy in Salmonidae – Living and moving in the fast lane. *North American Journal of Fisheries Management*, **17**, 1029–45.

O'Connor, W.C.K. & Andrew, T.E. (1998) The effects of siltation on Atlantic salmon, *Salmo salar* L., embryos in the River Bush. *Fisheries Management and Ecology*, **5**, 593–401.

O'Grady, M.F., King, J.J. & Curtin, J. (1991) The effectiveness of two physical in-stream works programmes in enhancing salmonid stocks in a drained lowland river system. In: *Strategies for the rehabilitation of salmon rivers* (ed. D. Mills). Atlantic Salmon Trust, Institute of Fisheries Management, Linnaen Society, London, 154–78.

Olsson, T.I. & Persson, B-G. (1986) Effects of gravel size and peat material concentrations on embryo survival and alevin emergence of brown trout, *Salmo trutta L., Hydrobiologia*, **135**, 9–14.

Olsson, T.I. & Persson, B-G. (1988) Effects of deposited sand on ova survival and alevin emergence in brown trout (*Salmo trutta* L.). *Archiv für Hydrobiologie*, **113**, 621–7.

Ormerod, S.J., Donald, A.P. & Brown, S.J. (1989) The influence of plantation forestry on the pH and aluminium concentration of upland Welsh streams: a re-examination. *Environmental Pollution*, **62**, 47–62.

Osborne, L.L., Bayley, P.B., Higler, L.W.G., Statzner, B., Triska, F. & Ivensen, M.T. (1993) Restoration of lowland streams: an introduction. *Freshwater Biology*, **29**, 187–94.

Österdahl, L. (1969) The smolt run of a small Swedish river. *Swedish Salmon Research Institute*, **8**, 205–15.

Ottaway, E.M. & Clarke, A. (1981) A preliminary investigation into the vulnerability of young trout (*Salmo trutta* L.) and Atlantic salmon (*S. salar* L.) to downstream displacement by high water velocities. *Journal of Fish Biology*, **23**, 221–7.

Ottaway, E.M., Carling, P.A., Clarke, A. & Reader, N.A. (1981) Observations on the structure of brown trout, *Salmo trutta* Linnaeus, redds. *Journal of Fish Biology*, **19**, 593–607.

Padmore, C.L. (1998) The role of physical biotopes in determining the conservation status and flow requirements of British rivers. *Aquatic Ecosystem Health & Management*, **1**, 25–35.

Parkinson, E.A. & Slaney, P.A. (1975) A review of enhancement techniques, applicable to anadromous game fishes. *Fish Management Report British Columbia Fish and Wildlife Branch*, **66**, 1–100.

Patrick, S., Waters, D., Juggins, S. & Jenkins, A. (1991) *The United Kingdom Acid Waters Monitoring Network. Site descriptions and methodology report*. Ensis, London, 1–63.

Pawson, M.G. (1982) Recapture rates of trout in a 'put and take' fishery: Analysis and management implications. *Fisheries Management*, **13**, 19–32.

Payne, R.H., Child, A.R. & Forrest, A. (1972) The existence of natural hybrids between the European trout and the Atlantic salmon. *Journal of Fish Biology*, **4**, 233–6.

Pemberton, R.A. (1976a) Sea trout in North Argyll lochs, population, distribution and movements. *Journal of Fish Biology*, **9**, 157–79.

Pemberton, R.A. (1976b) Sea trout in North Argyll sea lochs: 2. Diet. *Journal of Fish Biology*, **9**, 195–208.

Penaz, M., Kubicek, F., Marvan, P. & Zelinka, M. (1968) Influence of the Vir River Valley reservoir on the hydrobiological and ichthyological conditions in the River Svratka. *Acta Scientiarum Naturalium Academiae, Academie Scientiarum Bohemoslovacae Brno*, **2**, 1–60.

Pentelow, F.T.K., Southgate, B.A. & Bassindale, B. (1933) The proportion of the sexes and the food of smolts of salmon (*Salmo salar* L.) and sea trout (*Salmo trutta* L.) in the Tees estuary. *Fishery Investigations Ministry of Agriculture & Fisheries, Series 1, 3*, **4**, 1–10, London.

Peter, A. (1995) Untersuchungen über die Konkurrenz zwischen Bach- und Regenbodenforellen: Beispiele aus dem Einzugsgebiet der Bodenseezuflüsse. *Bundesamt für Umwelt, Wald und Landschaft. Mitteilungen zur Fischerei*, **55**, 89–108.

Peterson, R.H. (1978) Physical characteristics of Atlantic salmon spawning gravel in some New Brunswick streams. *Technical Report, Biological Station, St. Andrews, New Brunswick*, **111**, 1–28.

Peterson, R. H. & Metcalfe, J.L (1981) Emergence of Atlantic salmon fry from gravels of varying composition: A laboratory study. *Canadian Technical Report of Fisheries and Aquatic Sciences*, **1020**, 1–15.

Petts, G.E. (1984) *Impounded Rivers: Perspectives for ecological management*. John Wiley & Sons, Chichester.

Phillips, R.W., Lantz, R.L., Claire, E.W. & Moring, J.R. (1986) Some effects of gravel mixtures on emergence of coho salmon and steelhead trout fry. *Transactions of the American Fisheries Society*, **104**, 461–6.

Pirhonen, J., Koskela, J. & Jobling, M. (1997) Differences in feeding between 1+ and 2+ hatchery brown trout exposed to low water temperature. *Journal of Fish Biology*, **50**, 678–81.

Platts, W.S., Shirazi, M.A. & Lewis, D.H. (1979) Sediment particle sizes used by salmon for spawning with methods of evaluation. *Corvallis Environmental Research Laboratory, U.S. Environmental Pollution Agency, Corvallis, Oregon*. Report EPA-600/3-79-043, 1–32.

Pollard, R.A. (1955) Measuring seepage through salmon spawning gravel. *Journal of the Fisheries Research Board of Canada*, **12**, 706–41.

Potter, E.C.E. (1988) Movements of Atlantic salmon, *Salmo salar* L., in an estuary in south west England. *Journal of Fish Biology* **33 (A)**, 153–9.

Potter, E.C.E., Solomon, D.J. & Buckley, A.A. (1992) Estuarine movements of adult Atlantic salmon (*Salmo salar* L.) in Christchurch Harbour, southern England. In: *Remote Monitoring and Tracking of Animals* (eds I.G. Priede & S.M. Swift). Ellis Harwood, New York, 400–409.

Pratten, D.J. & Shearer, W.M. (1983) Sea trout of the North Esk. *Fishery Management*, **14**, 49–65.

Priede, I.G., Solbé, J. F., Nott, J.E., O'Grady, K.T. & Cragg-Hine, D. (1988) Behaviour of adult Atlantic salmon, *Salmo salar* L., in the estuary of the River Ribble in relation to variations in dissolved oxygen and tidal flow. *Journal of Fish Biology*, **33 (A)**, 133–9.

Pyefinch, K.A. (1955) A review of literature on the biology of the Atlantic salmon (*Salmo salar* L.). *Freshwater and Salmon Fisheries Research*, **9**, 1–24.

Pyefinch, K.A. & Mills, D H. (1963) Observations on the movements of Atlantic salmon (*Salmo salar* L.) in the River Conon and the River Meig, Ross-shire. I. *Freshwater and Salmon Fisheries Research*, **31**, 1–24.

Quinn, T.R. & Tallman, R.F. (1987) Seasonal environmental predictability and homing in riverine fishes. *Environmental Biology of Fishes*, **18**, 155–9.

Radford, A.F., Hatcher, A. & Whitmarsh, D. (1991) *An Economic evaluation of salmon fisheries in Great Britain*. Report prepared by University of Portsmouth Centre for the Economics and Management of Aquatic Resources for the Ministry of Agriculture, Fisheries and Food. Volume I, 1–292; II, 1–54; and Summary, 1–32.

Randall, R.G., Thorpe, J.E., Gibson, R.J. & Reddin, D.G. (1986) Biological factors affecting age at maturity in Atlantic salmon (*Salmo salar*). In: *Salmonid age at maturity* (ed. D.J. Meerburg). *Canadian Special Publications in Fisheries and Aquatic Sciences*, **89**, 90–6.

Rasmussen, G. (1986) The population dynamics of brown trout (*Salmo trutta* L.) in relation to year-class size. *Polskie Archiwum Hydrobiologii*, **33**, 489–508.

Raven, P.J., Holmes, N.T.H., Dawson, F.H., Fox, P.J.A., Everard, M., Fozzard, I.R. & Rouen, K.J. (1998) River Habitat Quality the physical character of rivers and streams in the UK and Isle of Man. *Environment Agency, Scottish Environment Protection Agency and Environment & Heritage Service, Northern Ireland*.

Raymond, H.L. (1979) Effects of dams and impoundments on migrations of juvenile chinook salmon and steelhead trout from the Snake River, 1966 to 1975. *Transactions of the American Fisheries Society*, **108**, 509–29.

Reisenbichler, R.R. (1997) Genetic factors contributing to declines of anadromous salmonids in the Pacific northwest. In: *Pacific salmon and their ecosystems* (eds D.J. Stouder, P.A. Bissen & R.J. Naiman). Chapman & Hall, New York, 223–44.

Ricker, W.E. (1954a) Pacific salmon for Atlantic waters. *Canadian Fish Culturist*, **16**, 6–14.

Ricker, W.E. (1954b) Stock and recruitment. *Journal of Fisheries Research Board of Canada*, **11**, 559–623.

Ringler, N.H. & Hall, J.D. (1975) Effects of logging on water temperature and dissolved oxygen in spawning beds. *Transactions of the American Fisheries Society*, **104**, 111–21.

Roberts, B.C. & White, R.G. (1992) Effects of angler wading on the survival of trout eggs and pre–emergence fry. *North American Journal of Fisheries Management*, **12**, 450–59.

Robinson, M. (1980) The effect of pre-afforestation drainage on the stream flow and water quality of a small upland catchment. *Institute of Hydrology Report*, **73**, 1–232.

Robinson, M. (1990) Impact of improved land drainage on river flows. *Institute of Hydrology Report*, **113**, 1–226.

Robinson, M., Jones, T.K., & Blackie, J.R. (1994) The Coalburn catchment experiment – 25 year review. *National Rivers Authority R & D Note*, **270**, 1–54.

Royal Commission on Environmental Pollution (1992a) *Freshwater Quality*. Sixteenth Report. HMSO, London 1–291.

Royal Commission on Environmental Pollution (1992b) *Freshwater Quality.* Additional Reports undertaken for the Royal Commission on Environmental Pollution. HMSO, London, 1–456.

Royal Commission on Environmental Pollution (1996) *Sustainable Use of Soil* – Nineteenth Report. HMSO, London, 1–209.

Rundquist, L.A., Bradley, N.E. & Jennings, T.R. (1986) Planning and design of fish stream rehabilitation. In: *5th Trout Stream Habitat Improvement Workshop* (eds J.A. Arway & R.F. Carline). Lock Haven University, 119–32.

Sadler, K. & Lynam, S. (1987) Some effects on the growth of brown trout from exposure to aluminium at different pH levels. *Journal of Fish Biology*, **31**, 209–19.

Sægrov, H., Hindar, K. & Urdal, K. (1996) Natural reproduction of anadromous rainbow trout in Norway. *Journal of Fish Biology*, **48**, 292–4.

Salmon Advisory Committee (1993) *Factors affecting emigrating smolts and returning adults.* MAFF Publications, London. 1–39.

Salmon Advisory Committee (1997) *Fish passes and screens for salmon.* MAFF Publications, London, 1–44.

Saunders, R.L. & Henderson, E.B.C. (1969) Survival and growth of Atlantic salmon fry in relation to salinity and diet. *Journal of the Fisheries Research Board of Canada Technical Paper*, **148**, 1–7.

Scott, A. & Beaumont, W.R.C. (1994) Improving the survival rates of Atlantic salmon (*Salmo salar* L.) embryos in a chalkstream. Institute of Fisheries Management. Cardiff (1993), 1–19.

Scottish Environment Protection Agency (1999) *Improving Scotland's Water Environment.* Scottish Environment Protection Agency.

Scottish Office Agriculture and Fisheries Department (1995) *Notes for guidance on the provision of fish passes and screens for the safe passage of salmon.* Scottish Office Agriculture and Fisheries Department.

Scottish Tourist Board & the Highlands and Islands Development Board (1989) *Economic*

Importance of salmon fishing and netting in Scotland. Report prepared by Mackay Consultants, Inverness.

Seber, G.A.F. (1973) *The Estimation of Animal Abundance.* Griffin, London.

Shchurov, I.L. & Shustov, Y.A. (1989) Comparison of the physical strength of juvenile salmon and trout in river conditions. *Journal of Ichthyology*, **29**, 161–3.

Shearer, W.M. (1992) *The Atlantic Salmon. Natural History, Exploitation and Future Management.* Fishing News Books, Oxford.

Shepherd, B.G., Hartman, G.F. & Wilson, W.J. (1986) Relationships between stream and intragravel temperatures in coastal drainages and some implications for fisheries workers. *Canadian Journal of Fisheries Science*, **43**, 1818–22.

Sheridan, W.L. (1962) Relation of stream temperature to timing of pink salmon experiments in southeast Alaska. In: *Symposium on pink salmon* (ed. N.J. Wilünovsky). H.R. MacMillan Lectures in Fisheries, U.B.C., Vancouver, 87–102.

Shirazi, M.A., Lewis, D.H. & Seim, W.K. (1981) Monitoring spawning gravel in managed forested watersheds. *Corvallis Environmental Research Laboratory, U.S. Environment Protection Agency, Corvallis Oregon* Report EPA-600/3-79-014, 1–13.

Slaney, P.A. & Zaldokas, D.O. (eds) (1997) *Fish Habitat Rehabilitation Procedures.* Watershed Restoration Technical Circular No. 9. Ministry of Environment, Lands and Parks, Vancouver.

Smirnov, A.I. (1955) The effect of mechanical agitation at different periods of development of eggs of autumn chum salmon (*Oncorhynchus keta* infrasp. *autumnali*s Berg. Salmonidae). *Doklady Akademii nauk SSSR*, **105**, 873–6 (Translated from Russian by Fisheries Research Board of Canada Translation Series 230, 1959.)

Smirnov, A.I. (1975) The biology, reproduction and development of Pacific salmon. *Isdaiya Moskva Universitet*, 1–335. (Translated from Russian by Fisheries and Marine Translation Service No. 3861, 1976.)

Smith, G.W., Smith, I.P. & Armstrong, S.M. (1994) The relationship between river flow and entry to the Aberdeenshire Dee by returning adult salmon. *Journal of Fish Biology*, **45**, 958–60.

Smith, I. & Lyle, A. (1979) Distribution of fresh waters in Great Britain. *Institute of Terrestrial Ecology*, 1–44.

Smith, I.P. & Smith, G.W. (1997) Tidal and diel timing of river entry by adult Atlantic salmon returning to the Aberdeenshire Dee, Scotland. *Journal of Fish Biology*, **50**, 463–74.

Smith, K. (1968) Some thermal characteristics of two rivers in the Pennine area of northern England. *Journal of Hydrology*, **6**, 54–65.

Smith, K. (1971) Some features of snow-melt recession in the upper Tees basin. *Water and Water Engineering*, 345–6.

Solbé, J. (1988) Water Quality for salmon and trout. *Atlantic Salmon Trust*, 1–58.

Solomon, D.J. (1978a) Some observations on salmon smolt migration in a chalk stream. *Journal of Fish Biology*, **12**, 571–4.

Solomon, D.J. (1978b) Migration of smolts of Atlantic salmon (*Salmo salar* L.) and sea trout (*Salmo trutta* L.) in a chalk stream. *Environmental Biology of Fish*, **3**, 223–9.

Solomon, D.J. (1982) Migration and dispersion of juvenile brown trout and sea trout. In: *Proceedings of the salmon and trout migratory behaviour symposium, Seattle.* (eds E.C. Brannon & E.O. Salo). School of Fisheries, University of Washington, 136–45.

Solomon, D.J. (1983) Salmonid enhancement in North America. *Atlantic Salmon Trust*, 1–40.

Solomon, D.J. (1985) Salmon stocking and recruitment and stock enhancement. *Journal of Fish Biology*, **27 (A)**, 45–57.

Solomon, D.J. & Child, A.R. (1978) Identification of juvenile natural hybrids between Atlantic salmon (*Salmo salar* L.) and trout (*Salmo trutta* L.). *Journal of Fish Biology*, **12**, 499–501.

Solomon, D.J. & Templeton, R.G. (1976) Movements of brown trout *Salmo trutta* L. in a chalk stream. *Journal of Fish Biology*, **9**, 411–23.

Somme, S. (1954) Undersökelser over laksems og sjöörretensgyting i Eira. *Jeger og Fisker, Oslo*, **6**, 7–10.

Somme, S. (1960) The effects of impoundment on salmon and sea trout rivers. *Natural aquatic resources. 7th Technical meeting IUCN International Union for the Conservation of Nature.* **Theme 1**, **4**, 77–80.

Spaas, J.T. (1960) Contribution to the comparative physiology and genetics of the European Salmonidae. III. Temperature resistance at different ages. *Hydrobiologia*, **15**, 78–88.

Stansfeld, J.R.W. (1989) Is a rod-caught salmon more valuable to the Scottish Economy than one caught in a net? *Salmon Net*, **21**, 59–64.

Stearley, R.F. (1992) Historical ecology of the Salmoninae. *Ecology of North American Freshwater Fishes.* Stamford University Press, Stamford, California, 622–58.

Stewart, L. (1969) Criteria for safeguarding fisheries, fish migration and angling in rivers. *Yearbook of the Association of River Authorities 1969*, 134–9.

Stocker, Z.S.J. & Williams, D.D. (1972) A freeze core method for describing the vertical distribution of sediments in a stream bed. *Limnology and Oceanography*, **17**, 136–8.

Stoner, J.H., Wade, K.R. & Gee, A. (1984) The effects of acidification on the ecology of streams in the upper Tywi catchment in west Wales. *Environmental Pollution*, Series A, **35**, 125–57.

Strevens, A.P. (1999) Impacts of groundwater on the trout fishery of the River Piddle, Dorset; and an approach to their alleviation. *Hydrological Processes*, **13**, 487–96.

Struthers, G. (1989) Salmon smolt migration from hydro–electric reservoirs. In: *Water Schemes: The Safeguarding of Fisheries* (ed. J. Gregory). Atlantic Salmon Trust, Pitlochry, 71–85.

Stuart, T.A. (1953a) Spawning migration, reproduction and young stages of loch trout (*Salmo trutta* L.). *Freshwater Salmon and Fisheries Research*, **5**, 1–39.

Stuart, T.A. (1953b) Water currents through permeable gravels and their significance to spawning salmonids. *Nature*, **172**, 407–8.

Stuart, T.A. (1957) The migration and homing behaviour of brown trout (*Salmo trutta* L.). *Freshwater and Salmon Fisheries Research*, **18**, 1–27.

Stuart, T.A. (1962) The leaping behaviour of salmon and trout at falls and obstructions. *Freshwater and Salmon Fisheries Research*, **28**, 1–46.

Sutcliffe, D.W. (1979) Some notes to authors on the presentation of accurate and precise measurements in quantitative studies. *Freshwater Biology*, **9**, 397–402.

Symons, P.E.K. (1971) Behavioural adjustment of population density to available food by juvenile Atlantic salmon. *Journal of Animal Ecology*, **40**, 569–88.

Symons, P.E.K. & Heland, M. (1978) Stream habitats and behavioural interactions of underyearling and yearling Atlantic salmon (*Salmo salar* L.). *Journal of the Fisheries Research Board of Canada*, **35**, 175–83.

Taccogna, G. & Munro, K. (1995) (eds) *The Streamkeepers' Handbook: a Practical Guide to Stream and Wetland Care*. Salmon Enhancement program, Department of Fisheries and Oceans, Vancouver, British Columbia.

Taylor, A.H. (1978) An analysis of the trout fishing at Eyebrook – a eutrophic reservoir. *Journal of Animal Ecology*, **47**, 407–23.

Thorpe, J.E. (1980) Ocean ranching. In: *Atlantic salmon: its future* (ed. A.E.J. Went). Fishing News Books Ltd, Farnham, Surrey, England, 2–14.

Thorpe, J.E. (1994) Significance of straying in salmonids and implications for ranching. *Aquaculture and Fisheries Management*, **25 (2)**, 183–90.

Turnpenny, A.W.H. (1989) Impoundment and abstraction – exclusion of fish from intakes. In: *Water Schemes. The safeguarding of fishes* (ed. J. Gregory). Atlantic Salmon Trust 87–111.

Valdimarsson, S.K. & Metcalfe, N.B. (1998) Shelter selection in juvenile Atlantic salmon, or why do salmon seek shelter in winter? *Journal of Fish Biology*, **52**, 42–9.

Vaux, W.G. (1962) Interchange of stream and intragravel water in a salmon spawning riffle. *Special Scientific Report, U.S. Fisheries and Wildlife Service*, **405**, 1–11.

Verspoor, E. (1988) Widespread hybridization between native Atlantic salmon, *Salmo salar*, and introduced brown trout, *S. trutta*, in eastern Newfoundland. *American Journal of Fisheries Management*, **7**, 91–105.

Vivash, R.M. (1989) Better drainage and better fisheries. In: *Water Schemes. The Safeguarding of Fisheries* (ed. J. Gregory). Atlantic Salmon Trust, 17–24.

Vivash, R., Ottosen, O., Janes, M. & Sørensen, H.V. (1998) Restoration of the Rivers Brede, Cole and Skerne: Joint Danish and British EU-LIFE demonstration project II – The river restoratiion works and other related practical aspects. *Aquatic Conservation Marine and Freshwater Ecosystems*, **8**, 197–208.

Waiwood, B.A. & Haya, K. (1983) Levels of chorionase activity during embryonic development of *Salmo salar* under acid conditions. *Bulletin of Environmental Contamination and Toxicology*, **30**, 511–15.

Walker, K.F., Hillman, T.J. & Williams, D.W. (1979) The effects of the impoundment of rivers: an Australian case study. *Verhandlungen der Internationalen Vereinigung für Theoretische und Angewandte Limnologie*, **20**, 1695–701.

Wallace, J.C. & Heggberget, T.G. (1988) Incubation of eggs of Atlantic salmon (*Salmo salar*) from different Norwegian streams at temperatures below 1°C. *Canadian Journal of Fisheries and Aquatic Sciences*, **45**, 193–6.

Ward, R.C. (1981) River systems and river regimes. In: *British Rivers* (ed. J. Lewin). George Allen & Unwin, 1–33.

Ward, J.V. (1976) Comparative limnology of differentially regulated sections of a Colorado mountain river. *Archiv für Hydrobiologie*, **78**, 319–42.

Watt, W.D. & Penney, G.H. (1980) Juvenile salmon survival in the Saint John River system. *Canadian Technical Report of the Fisheries Aquatic Sciences*, **939**, 1–13.

Weatherley, N.S., Campbell-Lendrum, E.W. & Ormerod, S.J. (1991) The growth of brown trout (*Salmo trutta*) in mild winters and summer droughts in upland Wales: model validation and preliminary predictions. *Freshwater Biology*, **26**, 121–31.

Weatherley, N.S. & Ormerod, S.J. (1990) Forests and the temperature of upland streams in Wales: a modelling exploration of the biological effects. *Freshwater Biology*, **24**, 109–22.

Webb, B.W. & Clark, E. (1998) Bed temperatures in Devon rivers, southern England, and some ecological implications. In: *Hydrology in a Changing Environment* (eds H. Wheater & C. Kirby). John Wiley & Sons, Chichester, 365–79.

Webb, B.W., Walling, D.E., Zhang, Y. & Clark, E. (1996) Water temperature behaviour in South-West England. *Hydrologie dans les payes celtiques*. INRA, Paris, 53–64.

Webb J. (1989) The movements of adult salmon in the River Tay. *Scottish Fisheries Research Report*, **44**, 1–32.

Whelan, K.F. (1991) An overview of techniques used in Atlantic salmon restoration and rehabiltation programmes. In: *Strategies for the rehabilitation of salmon rivers* (ed. D. Mills). Atlantic Salmon Trust, Institute of Fisheries Management, Linnaean Society, London, 6–18.

Whelan, B.J. & Marsh, G. (1988) *An Economic Evaluation of Irish Angling*. Report prepared for the Central Fisheries Board. The Economic and Social Research Institute, Dublin, 1–84.

White, T.A. (1942) Atlantic salmon redds and artificial spawning beds. *Journal of the Fisheries Research Board of Canada*, **6**, 37–44.

Wickett, W.P. (1952) Production of chum and pink salmon in a controlled stream. *Progress Report, Fisheries Research Board of Canada*, **93**, 7–9.

Wickett, W.P (1954) The oxygen supply to salmon eggs in spawning beds. *Journal of the Fisheries Research Board of Canada*, **11**, 933–53.

Willson, M.F. (1997) Variation in Salmonid life histories, patterns and perspectives. *United States Department of Agriculture, Forest Service. Pacific North West Research Station.* Research Paper . PNW-RP-498, 1–50.

Winstone, A.J. (1989) A review of techniques for salmon stock monitoring in relation to tidal power barrages. In: *Salmon Stock Monitoring*. Report to the Department of Energy by the Welsh Water Authority, 1–29.

Winstone, A.J, Gee, A.S. & Varallo, P.V. (1985) The assessment of flow characteristics at certain weirs, in relation to the upstream movement of migratory salmonids. *Journal of Fish Biology*, **27 (A)**, 75–83.

Witzel, L.D. & MacCrimmon, H.R. (1983) Embryo survival and alevin emergence of brook charr, *Salvelinus fontinalis*, and brown trout, *Salmo trutta*, relative to gravel composition. *Canadian Journal of Zoology*, **61**, 1783–92.

Worthington, E.B. (1940) Rainbows: a report on attempts to acclimatize rainbow trout in Britain. *Salmon and Trout Magazine*, **100**, 241–61.

Worthington, E.B. (1941) Rainbows: a report on attempts to acclimatize rainbow trout in Britain. *Salmon and Trout Magazine*, **101**, 62–98.

Author Index

Subject Index

(Excludes trout, salmon, salmonid, Salmonidae which are common throughout).